The Playful Brain

The Playful Brain
Venturing to the Limits of Neuroscience

Sergio Pellis and Vivien Pellis

ONEWORLD

OXFORD

A Oneworld Book

Published by Oneworld Publications 2009

ISBN 978-1-85168-632-2

Typeset by Jayvee, Trivandrum, India
Cover design by Mungo Designs
Printed and bound in Great Britain by
TJ International, Padstow

Oneworld Publications
185 Banbury Road
Oxford OX2 7AR
England
http://www.oneworld-publications.com

Mixed Sources
Product group from well-managed
forests and other controlled sources
www.fsc.org Cert no. SGS-COC-2482
© 1996 Forest Stewardship Council
FSC

Learn more about Oneworld. Join our mailing list to
find out about our latest titles and special offers at:

www.oneworld-publications.com

To our parents

CONTENTS

FOREWORD

Psychologists have recently realized that happiness, laughter, joy, affection, and related phenomena have been neglected and are now developing a new specialty termed "positive psychology." For too long they were putting most of their efforts into dealing with psychological and behavioral disorders: grief, trauma, jealousy, anger, violence, fear, pain, suffering, and loss. There is also a growing focus on psychological well-being, pleasure, and even happiness in non-human animals in laboratories, zoos, and even in nature itself. Play, broadly conceived, may be a major process underlying lives worth living.

Serge and Vivien Pellis end their marvelous volume by asserting that play is still a mystery, and one to be enjoyed. Certainly a world without mysteries, let alone pleasure, would be a dull one. But solving mysteries is not only what gives scientists their kicks, but is really what science is all about. For scientists, solving one set of challenges inevitably raises further mysteries to contemplate and investigate. Procedurally, this situation is comparable to those early video games, where finding the location of one treasure brought on the next level and setting, which one had to puzzle out, typically for an even more valuable cache, all the while confronted with more dangerous pitfalls, traps, enemies, and competitors. This book raises the prospect of valuable treasures awaiting students of play, though not necessarily of the kind anticipated by those who pioneered the study of play in animals and children in the nineteenth and early twentieth centuries.

A careful reading of this book will show that social play, in particular, is far less of a mystery than it was even a decade ago. This advance is due in large part to the breadth and depth of the studies by Serge and Vivien Pellis, often in collaboration with many students and colleagues. Along with a comprehensive treatment of an often disparate and scattered literature, they have brought together in one compact volume the dozens of careful, insightful, elegant, even brilliant studies they and their colleagues have contributed over thirty years. In so doing the seminal nature of their contributions are made evident.

What are the treasures this book contains? Readers will need to uncover their own nuggets, sometimes scraping through the modest overlay in which Serge and Vivien embed them. These nuggets need to be amalgamated, however, to appreciate how these rich empirical data not only extend, but actually refocus, our understanding of how play evolves across species, develops in individuals, and functions in real life. They do this by carefully describing the details of play in many species and analyzing the neural, hormonal, and other physiological and sensory systems underlying play.

The book covers many topics, often rather technical, yet remains accessible for a broad readership. It is filled with subtle humor and obvious passion on a topic rarely accorded systematic scientific study, with the possible exception of play in children. Educators, therapists, anthropologists, sociologists, recreation supervisors, and psychologists studying play have ignored the vigorous rough-and-tumble play focused on here, while biologists and ethologists rarely connect their studies on play in non-human animals with the rich literature on children. On the other hand, neuroscientists and geneticists are beginning to tap into the tapestry of play to access processes of neural organization, genetics, affect and cognition, and sensory – motor integration.

In recent years more books are appearing on play of all kinds, especially in children, apes, and monkeys. While these contributions are important, this book mines deeper into the recesses of animal behavior to emerge with its treasures. The initial mother lode was gleaned from the behavior of laboratory rodents: cheap, small, short-lived, and the repository of the most advanced and detailed physiological, genetic, behavioral, and neurological information available for any mammal. It is no accident that comparisons of lab rats and lab mice dominate chapters 2 and 3 and reappear frequently. But the twists and turns taken as the other rodents, monkeys, apes, humans, dogs, dolphins, and other species appear, strut their stuff, and fade away is reminiscent of the process of play itself. Does it take a playful mind to effectively accomplish the integration of so many disparate elements? Play as a calibrating mechanism for emotions, motor control, stress reduction, role relationships, and other important functions is strongly documented, while the usual obvious suspects for why animals play have proven increasingly difficult to support. In some cases nature works in a more complex and variable fashion than anticipated by our love of parsimonious, even simple, mechanisms.

It has long been recognized that play involves positive emotions and feelings as well as intense behavioral performances. It is engaging to watch animals play because it is variable: the variations on a theme format is a major staple of composers for good reason. The Pellises show how and why this variability is generated and how it affects the performers, not just the spectators. Furthermore, they show how a playful brain could have evolved. Here mouse–rat and other rodent comparisons are key routes to insights applicable to other species, including humans. Their success is due to their detailed comparative study of many species, from birds to apes, rather than the extrapolation from rodents to human beings that was the hallmark of an earlier experimental psychology. Social play turns out to be a phenomenon with both a deep and shallow evolutionary history. That is, while play fighting may go back in evolutionary time to fish and amphibians, its manifestation in monkeys, rodents, canids, seals, ungulates, and other groups is highly variable. This can only mean that play has functions in animal lives that vary even among closely related species. What is most exciting, however, is that we now have good data on the details of the differences, some of the neural and adaptive mechanisms involved in them, and the power of modern methods of phylogenetic analysis to organize the findings and even make predictions and develop applications, even in our own species.

This book deserves serious consideration by scientists and educators from all areas of biological and social sciences, including evolutionary biologists. Fortunately this book is just an installment of what is to come in the study of play and related phenomena over the next decade. The Pellis lab will continue to produce seminal empirical studies, descriptive as well as experimental, on play in many other species. An equally important effect of this book is to challenge those studying play to refine their approaches and theories, devise better explanations for functions of play, and consider the need for an integrative approach. I also suspect that persons reading this book will be inspired to join in the study of this most fascinating topic.

Gordon M. Burghardt

PREFACE

Play is an endlessly interesting subject. Over time, those interested in explaining its occurrence have generated more opinions than solid evidence about what it is, how it arises, and what it does for those who play. Indeed, in the history of the Western world, play has vacillated between being seen as instrumental to the development of healthy individuals and as a childish waste of time. Similarly, cross-cultural studies show that play by children can be valued, merely tolerated, or actively discouraged.[1] However, even when seen as valuable, it has not been treated as intrinsically so but as a means to an end – such as an educational tool, by which teachers, who cunningly use it as a medium, can manipulate students.[2] Play, whether seen as a valuable developmental tool or as a worthless, childish endeavor, was readily absorbed into psychology as an object of study when that discipline arose in the late nineteenth and early twentieth centuries. Although there is still much debate among psychologists concerning the development and function of play in children,[3] textbooks have been written explicitly for university courses on the subject.[4] However, the study of play that extends beyond humans to wider swathes of the animal kingdom has had a more troubled history.

Although Charles Darwin was a big influence in legitimizing the study of behavior as a meaningful biological property of organisms, it was not until 1898, when Karl Groos published his treatise,[5] that scientists turned their attention to the play of non-human animals. Unfortunately, because the field of animal behavior did not fully emerge as an independent, academic discipline until the middle of the twentieth century, much of the research on non-human animal play was sporadic, and many of the reports on the subject were buried in broader behavioral and ecological studies of animals. Despite this problem, research on non-human animal play began to pick up steam in the 1970s and became a legitimate problem worthy of broad attention in the early 1980s, as a result of the publication of two seminal works. The first, written by Bob Fagen, provided two services. By scouring the literature, and incorporating into his book studies written in both English and in other languages, he summarized what was known about the play of a wide range of species. Fagen then applied the theoretically rigorous methods of the evolutionary approach, which had, by then, become the dominant approach in animal behavior, to explore the functions that such play may serve during development. The second work was a review article by Peter Smith, which was not only a rigorous analysis of the function of play, but was also an attempt to integrate human and non-human play into a comprehensive, theoretical framework.[6] These writings influenced a generation of researchers in how they approached the study of non-human animal play, especially those trying to understand this behavior in free-living animals.[7]

In seeming independence of these trends in seeking the evolutionary functions of play, other researchers, such as Dorothy Einon in the UK and Jaak Panksepp in the US, coming from different traditions in experimental psychology, comparative psychology, and the newly emerging behavioral neuroscience, began to use laboratory animals – especially rats – to study the development of play, its influences on the emergence of cognitive and sexual behavior, and its neural underpinnings.[8] By the late twentieth and early twenty-first century, the number of scientific papers on the play of rats, especially on its development and its neural mechanisms, had greatly expanded.[9] Unfortunately, few researchers attempted to bring together these two approaches to the study of non-human animal play – the former, which focuses on the consequences of playing that led to its evolution, and the latter, which delves into ever greater detail on the nuts and bolts mechanisms that generate it. The lack of success by the first approach in identifying an empirically, well-supported function for play and the accumulation of interesting, but disparate facts, by the second approach, led to the waxing and waning of interest by researchers in studying non-human animal play. Despite this, as many young animals have the annoying habit of making play a major part of their life experience, attention is always drawn back to the issue.[10] Thus, play remains a problem that needs an explanation, and one way in which its study can progress is to combine the approaches that emphasize function with those that emphasize mechanism. What is needed is an overarching theoretical framework for play that can embrace them both.

One such approach is to look at play from the perspective of its origins. What were the conditions that led some animals to evolve play from ancestors that did not play? A champion of this historical perspective has been Gordon Burghardt, a researcher with a foot in both the experimental psychology and animal behavior traditions.[11] The historical perspective involves two steps. First, it requires that one not only study the obvious cases of play – those clear-cut cases in which most observers would agree that what they are seeing is play – but also the borderline cases – those in which one scratches one's head to wonder whether what one is observing is play or not. It is the incipient cases of play that enable the observer to identify the conditions that make play possible, since the difference between these cases and those species that do not play is small. Second, this perspective recognizes that when looking at species that do play, the behavior can differ in its degree of complexity, from rudimentary to elaborate. By examining the range of variation across species that do play, we can characterize the neural and behavioral elements that form its building blocks. This approach integrates knowledge about the mechanisms that generate play and the functions that play may serve, and does so by taking the diversity of play present in the animal kingdom seriously. We have utilized this integrative, historical approach, especially its second component, throughout most of our empirical work on non-human animal play and it is the unifying framework that we have adopted in this book.[12]

For whom have we written this book? After all, the organization of a book is usually structured according to its expected audience. But we have had several audiences in mind when writing this book: this has necessitated making compromises, and so we have run the risk that none of these potential audiences is satisfied. So let us explain who we think would find this book of interest.

The impetus for writing this book came when one of us, Sergio, gave a seminar to a rather eclectic audience – of anthropologists, educators, sociologists, psychiatrists, and developmental psychologists – on the lessons that we have learned about play from studies done on laboratory rats. The audience found that this research revealed much about play that was interesting and laden with potentially important implications for the study of humans and their psychological development. Our impression was confirmed by the response that we received from a short paper that we subsequently wrote on the topic for circulation to a wide audience that included educated lay readers.[13] From these reactions, we concluded that there could be a wide audience of people, of professionals, who deal with issues related to human development, and of lay readers, who, as parents or simply interested individuals, wish to learn about what those of us, from the various fields of animal behavior, comparative psychology, and behavioral neuroscience, who are studying play, have to say about this fascinating subject. For these readers, we kept the narrative as simple and free of jargon as possible. Where technical concepts and facts cannot be avoided, we use examples either from the literature or fictional scenarios to help make them vivid and clear. To simplify the text, we provide extensive endnotes, to offer literature to which the interested reader can go if they wish to explore topics further, or to elaborate on some issue that may need further explanation, but would be tangential if incorporated into the main body of the text.

Readers should also be aware that this book is not just a summary of what is known. Rather, we utilize our own journey in empirical research on play as a guiding framework within which to incorporate and evaluate what has been uncovered by dozens, if not hundreds, of researchers around the world. Thus, we view this book as a synthesis of what we have learned personally, and have offered here many new insights about how play originated, evolved, and has come to have the properties it does in some marvelous players such as rats and humans. We also wrote this book with our colleagues in animal behavior, comparative psychology, and behavioral neuroscience in mind. We hope that we present sufficient detail of our insights and the logic and data supporting them, for these readers to have the information necessary to challenge the veracity and strength of our conclusions by conducting more research. Naturally, we hope that some of the novel insights offered in this book will stand up to vigorous testing, but many may not. If this book can spur further work in this field that in the future leaves us knowing more than we do now, then the effort will have been worthwhile.

Another important audience that we had in mind while writing this book was students, who may be confronted with similarly complex behavior and are wondering how to tackle it. For these readers, we hope that our multi-decade experience, which is reflected in this book, offers them a guide for how to decompose a complex behavior and then find the rules by which to re-synthesize it. Juxtaposing neural mechanisms with behavioral mechanisms and embedding these within transformations over time, be it in the individual's lifetime or in transgenerational changes, provides a powerful working framework. We hope that this book will make these steps in the analysis clear to our successors, whether they are interested in play or in some other, complex behavior. Of course, we will always have a soft spot for students who decide that studying play is just too exciting to forego. If reading this book sparks such excitement, then again, we will be very satisfied with our efforts. Finally, we believe that play is fascinating, and we hope that most readers, whatever else they may gain from this book, will also simply enjoy the journey as it unfolds.

ACKNOWLEDGMENTS

Our journey in trying to understand play began with Sergio's graduate studies; he was joined nearly from the outset by Vivien. Over such a long period it becomes almost impossible to mention all those who directly or indirectly mentored, facilitated, collaborated, or encouraged our work, so we apologize in advance if, in the following, we have missed some names. Two direct mentors were John Nelson and Philip Teitelbaum, from whom we learned how to think about behavior and its relationship to the brain. Others, such as John Baldwin, Bob and Caroline Blanchard, Marc Bekoff, Gordon Burghardt, Stella Crossley, Mike Cullen, Don Dewsbury, Doug Dorward, Bob Fagen, John Fentress, Ilan Golani, Bryan Kolb, Alan Lill, Dan McIntyre, Gerlinde Metz, Gail Michener, Dennis O'Brien, Don Symons, Bernard Thierry, Ian Whishaw, and Dave Wolgin, provided invaluable collaborations, training, advice, encouragement, and challenging feedback. There are also innumerable colleagues whose work on play inspired our own efforts. There are too many to mention, but three who did critical work on the play of rats and that we came to know personally are Dorothy Einon, Jaak Panksepp, and Steve Siviy. Of course, a quick glance of citations to our own work reveals that a lot of what we have found could not have been done without the hard work, diligence and new ideas of the many students and post-doctoral fellows who have worked with us, such as Cheryl Arelis, Heather Bell, Karen Dean, Su-Lin Fantella, Evelyn Field, Margaret Forgie, Afra Foroud, Berna Dean Gaudry, Erica Hastings, Andrew Iwaniuk, Holly Kamitakahara, Joanna Komorowska, Neala MacDonald, Jo Manning, Mario McKenna, Marie Monfils, Eugenia Natoli, Tamara Pasztor, John Pierce, Christine Reinhart, Alan Salo, Takeshi

Shimizu, and Lori Smith. There are so many others not named whose efforts we appreciate. Research takes money and since coming to North America from our Australian homeland, two agencies have been critical in enabling us to maintain and develop our research program on play – the Harry Frank Guggenheim Foundation and the Natural Sciences and Engineering Research Council of Canada. The core of this book was written while Sergio was on a one-year study leave, and so the important institutional support provided by our employer, the University of Lethbridge, needs to be thanked in its continuing support of scholarly activity in general, and ours in particular.

But there is more than collecting data and feeding the intellect that goes into such an enterprise – there is also feeding the body. In the early stages of this journey, it was the unstinting financial and moral support of our parents, George and Doris Cosopodiotis and Marcello and Arles Pellis, and nanna, Maude Cosopodiotis, which kept us going through some bleak times. This is not to say we haven't received intellectual support from our families, we have. The enthusiasm for our efforts expressed to us by our aunts, uncles, siblings, nieces, nephews, and cousins has helped sustain our morale over the years. Education, money, encouragement, and growing CVs are all fine, but to write a book there needs to be an audience. That there is a potential audience for the ideas expressed in this book was the insight of Juliet Mabey from Oneworld Publications, and we are most grateful for her encouragement and guidance in getting this writing project started, and Fiona Slater and the other staff from Oneworld Publications for guiding us to the finish line. Finally, some of our colleagues, students, family and friends have been asked for more – to read chunks, or in some cases, the whole book, and provide feedback. For this we are most grateful, and hopefully, our worst errors or lapses of logic have been expunged due to their efforts and some of the artwork is better due to their input. We are especially thankful for the efforts of Louise Barrett, Gordon Burghardt, Karen Dean, Bob Fagen, Andrew Iwaniuk, Bryan Kolb, Susan Lingle, an anonymous reviewer, and to the students of the senior undergraduate course on play taught by Sergio in 2008. But, in the end, all the help in the world cannot absolve us of our responsibility for what has made it through to the printed page.

ILLUSTRATIONS

one on the right side of the long axis of the brain. At the rear of the brain is another large structure, the cerebellum, and coursing further to the rear is the spinal cord. Note that at the posterior of the brain, where the two cerebral cortices meet at the centre, there is a small triangular space and two mounds can be seen – these are the superior colliculi, brainstem structures that process visual information. In mammals with a larger cortex, such as humans, the cortices expand to cover these subcortical structures. At the front end of the brain, sitting above the olfactory lobes, is the prefrontal cortex, a structure that looms large in our story. Another area important to our tale, the motor cortex, is in the area just behind and to the right of the first vertical groove (Y-shaped line). In the panel below, and to the left, a sagittal section (length-wise along the middle of the brain) of a rat brain is shown with most of the various areas that are discussed in the text labelled, thus giving the reader an idea of the general layout of the mammalian brain. In the panel below, and to the right, is a coronal section (transverse across the brain) at the level of the prefrontal cortex, showing two areas of this part of the cortex that are discussed at length in the text: the orbital frontal cortex (OFC) and the medial prefrontal cortex (mPFC).   45

3.7. These graphs show that when compared across mammalian orders, there is a positive relationship between the prevalence of play and relative brain size. Note that this relationship holds for both raw data (A) and data corrected for the degree of relationship between orders (i.e. Independent Contrasts) (B). (From Iwaniuk, A. N., Nelson, J. E., & Pellis, S. M. (2001). Do big-brained animals play more? Comparative analyses of play and relative brain size in mammals. *Journal of Comparative Psychology,* **115**, 29–41. © 2001, APA, reprinted with permission.)   53

3.8. When the proportion of adult weight gained prior to birth is used, a significant negative correlation (see r-values above the graphs) is present whether a complexity score involving all facets of play fighting (see Table 3.1) is used (A), or one that simply scores whether species have a wrestling form of play fighting (B). That is, species that are more mature at birth play less. Conversely, this means that the greater the period of time spent as juveniles, the greater the

THE PUZZLE OF PLAY

Mbundi runs up to Ntondo from behind. As he runs past, he grabs Ntondo by the arm. Ntondo jerks his body back, and braces against being pulled forward. This brings Mbundi to a standstill, facing Ntondo. After looking at each other for a moment, Ntondo turns his head and lunges, with his mouth wide open, and bites Mbundi's hand, which is still holding Ntondo's arm. As Mbundi releases his hand and withdraws it, he lunges forward, grabs Ntondo by both shoulders and bites at the side of his neck. Ntondo ducks and rolls onto his side, Mbundi follows and falls on top of Ntondo. They grapple with one another, delivering gentle bites wherever they can. Eventually, Mbundi jumps up and begins to run away, but then slows down, and, with a wide-open mouth, looks back over his shoulder. At first, Ntondo is slow to stand up, but he then jumps to his feet, and chases after Mbundi.[1]

The preceding description of what many observers would call "rough-and-tumble play" or "play fighting," in two juvenile gorillas, could fit any number of primates, or with a reduction in the role of the hands, almost any mammal. Roughhousing play such as this is familiar to most readers through their children, recollections of their own childhood in the school-yard, pet cats and dogs that they may have reared, or from monkeys and apes seen at the zoo. But what is it that is playful about this behavior?

Play fighting is but one form of play, a behavior that can also involve objects, such as a kitten playing with a ball of wool, or one that is self-directed, such as a calf gamboling in a meadow. If we consider human children, play can also involve playing peek-a-boo or donning a superhero costume and fantasizing. Therefore, the range of behavior that can be labeled as play is quite extensive, especially when considering humans.[2] So what are the characteristics that define behavior as play? Historically, defining play has not been an easy task, and there is no single agreed-upon definition. Nevertheless, there are criteria that most researchers generally agree to be necessary components of a definition of play: it is an activity that is engaged in voluntarily, and it is positively reinforcing – that is, the performers find it pleasurable. Although these two criteria are true of other behaviors, such as eating and sex, another commonly incorporated criterion is that the purpose of play is not immediately utilitarian. This feature of play is often critical, since for non-human animals, play is often a simulation of a functional activity, such as fighting or predation. Thus, it is the absence of the normal consequences of these behaviors – the killing and eating of prey or the taking of food or some other resource from a social partner – that alerts the observer to the possibility that the behavior being observed is play.

There is an overabundance of definitions posited for play, with much of the confusion and argument arising from the different perspectives of various authors. For example, some authors have focused on the properties of play behavior, such as its seemingly exaggerated movements, while others have focused on the outcomes of the behavior, such as whether it simulates sex or aggression. Some definitions that emphasize a specific property of play, such as its lack of immediate function, have been the most widely adopted, but they run the risk of being so general that most behaviors that are performed by immature animals could result in being characterized as play. Therefore, in most cases, we rely on our intuition to distinguish between play and non-play.[3] However, the less like humans the species being observed are, the greater the likelihood that our intuition may lead us astray.

For example, are two immature cockroaches pushing and lashing each other with their antennae, for what seems like no apparent purpose, exhibiting play? Unlike the case of two puppies play fighting, we are not privy to a body language that makes sense to us – for puppies, wagging tails, floppy ears, and a relaxed body tone are most often the clues that we use to distinguish such behavior from serious aggression. We use similar clues from body language when trying to decide whether two children are playing or fighting. When videotaped sequences of play fighting and serious fighting are shown to children and to adults, both use a variety of clues ranging from whole body movements to facial expressions in order to be able to differentiate between the two behaviors. Because there are multiple cues that can be used to distinguish between playful and serious fighting, if faces are not visible, other features of the body movements can instead be used. Obviously, the more acquainted observers are with the species in question, the more likely they are to have access to a range of criteria, whereas the more alien the species, the more limited the criteria. In addition, when studying human play, subjects that are engaging in either playful or serious fighting can, of course, be asked as to whether their behavior was playful or serious, so as to verify the judgments of the observers that are based on the behavioral cues alone.[4] However, even this last item is not foolproof, as humans are notorious for telling untruths: after all, how often do we smile when we are really not happy? Nonetheless, when trying to decide whether the behavior of species other than human should be labeled as play, no such additional information, however suspect, is available. Yet, in order to comprehend the problem of play fully, we need to know which borderline cases in the animal kingdom belong to the play category and which do not.

Developing a more exhaustive range of criteria that can readily distinguish between the playful and non-playful behavior of all animals, be they immature or adult, has two advantages: it brings the non-human literature into closer correspondence with the human literature, and it enables us to compare a diverse range of animals. For example, Gordon Burghardt has developed a set of criteria that must be met for an instance of behavior to qualify as play: (1) that the behavior is incompletely functional in the context expressed; (2) that it is voluntary; (3) that it is, in some ways, structurally modified or temporally shifted as compared to when it is used in its normal, functional context; (4) that it is performed repeatedly, but not necessarily in an invariant form; and (5) that it is present in healthy, unstressed animals. In using these five criteria, researchers have shown that animals as diverse as turtles, wasps, and octopuses engage in behavioral sequences that are comparable to those performed by mammals such as dogs and monkeys – ones to which most observers would happily apply the label of play.[5] Unfortunately, in most cases, the lack of detailed data means that comprehensive comparisons across the entire animal kingdom so as to determine the presence of play limit our attempts to plausible best guesses.

However, if we just focus on the cases where most observers would agree that what they are seeing is play, we are still left with an interesting phenomenon – that the majority of mammals and many birds spend some of their time in a seemingly pointless behavior. Based on studies of both free-living and captive animals, play occupies up to 20% of an animal's daily time budget and up to 10% of their daily energy budget. Given that these figures are for animals in their juvenile period, this implies that these individuals are sacrificing resources that could be channeled into growth. Furthermore, when playing, animals risk making themselves conspicuous to predators and chance injury. Although there is debate about how significant a burden these costs may actually be, there is little doubt that play involves some cost, such as spending less time feeding, and so must also have some compensating benefits – even if these benefits are modest in magnitude.[6]

Commensurate with the breadth of human imagination, there is a long list of proposed benefits of play. Most are predicated on the view that, because play has no immediate purpose, the benefits are delayed. That is, juveniles play so as to become better adults. Occasionally, scholars have been drawn to the possibility that play may serve to build a better juvenile[7] – but this would then imply an immediate function, which creates more problems for the definition of play. To make matters worse, some have posited that in most cases, what is labeled play serves no function whatsoever, be it delayed or immediate. If play is a property of immaturity, then how immature organisms attempt to perform actions may, from an adult observer's point of view, appear playful. In this scenario, play is an epiphenomenon of processes enacted in the immature stages of development, and it is these processes that need explanation, not the incidental by-products of such processes.[8] There is good reason why these views sound confusing and even contradictory to us. Like all biological traits, play has evolved, and evolution is a messy process. The origin of traits need not reflect their current functions, and traits used for one function at a particular time may be co-opted for another function at another time. For example, two adaptations associated with modern birds are long, hollow bones and wings that use feathers as an aerofoil. Yet the origins of feathers pre-date birds – dinosaur ancestors likely used feathers for insulation, not flight, and hollow limb bones seem to have pre-dated even dinosaurs with feathers.[9] This is also likely to be true for play.

There is no doubt that many properties of animals (such as being "hot-blooded" or "cold-blooded"), and developmental processes (such as the piecemeal maturation of sensory and motor capacities), are, when brought together in a particular combination, likely to generate play-like behavior. Such play-like behavior may be functionless, but once present, it could then be co-opted for some useful function, with those features that are the most important to that function being honed to serve that function better. At some future point, that modified play behavior may again be co-opted for another function and so undergo more changes appropriate for the new function. Some lineages of currently playful organisms may reflect any or all of these levels of transformation.[10] It is this complex tapestry created by millions of animals evolving over vast periods of time that causes us to stumble when we try to encapsulate these diverse patterns and processes into a single definition or explanatory theory of play. But this very richness in variation is also an advantage. Nature has provided us with a vast array of natural experiments, with different lineages of organisms having changed play in this way or that, which allow us to exploit this variability as a useful tool for research.

However, the downside of using natural diversity is in making sure that what is being compared is indeed comparable. For example, if species A plays in a complex manner with sticks or other inanimate objects, and species B does so with other members of its species, is the play comparable? Even if we only compare play that occurs between members of the same species, can we be sure that the play-like fighting in cockroaches bears any relation to that of gorillas? Our solution to this problem is to focus attention on just one form of play, and to do so by examining how it varies in just one lineage of related animals. In the comparative literature, play fighting, or rough-and-tumble play, is the most commonly reported form of play. And, within that literature, the most detailed and extensive experimental research has focused on laboratory rodents, with the rat being especially prominent.[11] Furthermore, there is sufficient detail available on a dozen or so rodents for us to determine what features of play fighting are unique to rats and what features of play fighting are shared across several species of rodents. A comparative analysis allows us to identify those features of play fighting that may be generalized to all species that engage in play fighting. What we learn from rats can then be applied to species closer to our hearts and minds, such as ourselves. In addition, the approach advocated here is informative in another way.

Comparisons among closely related species can provide clues as to what features can be changed independently of one another and how those changes may be related to the novel functions that play may have for that species. Also, the identification of those changes can lead to useful hypotheses about the accompanying brain changes needed to enact such a behavioral change. So, the first task of the comparative analysis that we use here is to chart the pathway by which a particular species has come to acquire its unique characteristics of play fighting.[12]

There is also another benefit in using the comparative approach. Once the unique features of a trait are identified in some species in one lineage of related species, the net can be cast more widely to see if other lineages have sprouted species with similar characteristics. Of all of the rodent species that have been studied, rats have evolved the most complex pattern of play fighting, with control mechanisms over play that resemble those seen in other lineages, such as primates. In some features, the play fighting of rats closely resembles that which is present in humans and in related non-human primates.[13] When traits in species from two different lineages converge in such a manner, it is likely that those features have evolved in both of them to solve similar problems. This convergence offers two advantages to the researcher. First, we can use the comparison to help identify what the common social or environmental problems may be that facilitate the evolution of such traits. Second, we can identify species that may be good models for the experimental analysis of that trait. As we shall see, because the play fighting of rats shares at least some features in common with the play fighting of humans, we have an avenue available to study those traits in a way that cannot be done with human subjects.

Our laboratory has made two primary contributions to this endeavor. First, we examined, in detail, the play fighting in rats and many other rodents and characterized the key components of this behavior (see above). Second, in a series of experiments, we showed that the cortex, the most evolutionarily advanced part of the mammalian brain, is not necessary for rats to engage in play fighting. That is, rats without a cortex (decorticate) are still able to use all the behavior patterns typical of play fighting and to engage in such play just as frequently as rats with intact brains. Consequently, decorticate rats not only want to play, but also know how to play. However, when they play, their behavior is not without abnormalities. These rats fail to modify how they play at different ages; rather, at all ages, they play in the same manner. Nor do they modify their play when interacting with different partners – they play in the same manner whether their partner is a dominant male, a subordinate male, or a female. Furthermore, these age-related and partner-related modifications in play fighting involve different neural circuits in the cortex, as we will explain in greater detail in a later chapter. Thus, there are distinct cortical and subcortical controls over play fighting in rodents, with some species having all the control mechanisms and others having few or even none. Based on this framework, we can use the comparative data set to ask the question: "How can a brain, capable of producing play fighting, be constructed from one that cannot?"

A Cause for Rejoicing and A Cause for Caution

The rat thus provides a tool by which to chart the evolution of play fighting. As there is more known about the brain and behavior of the laboratory rat than virtually any other mammal,[14] there is a solid biological framework within which to chart the evolutionary transformation that has led to the pattern of play now seen in the modern-day rat. For these reasons, we think that focusing both on rats and on play fighting – the most frequent form of play in rats – is a very useful way to advance our understanding of play in general. There are limitations, of course, but even with those limitations, we believe that the picture provided by the rat when placed within a broader, comparative context provides, at the very least, a framework that can be used as a model for many other forms of play as well as for other species, including humans.

Whereas rodents are but one order of the twenty orders of mammals (e.g. carnivores, cetaceans, marsupials, primates, etc.), they comprise over 40% of the 4500 or so mammalian species still in existence. So, it remains uncertain how representative the play described in the couple of dozen species that have been studied is for the order as a whole. Furthermore, since the majority of those studied are small (rat-sized or smaller) and mostly nocturnal, there are relatively few data available on social and non-social behavior under free-living conditions.[15] Thus, while we know quite a lot about laboratory rats, laboratory mice, and a few other rodents, our knowledge of these species under more natural conditions is limited. And for diurnal species that live in the open, such as ground squirrels, details of play are available for only a handful.[16] In addition to these concerns is the fact that the principal subject in our story, the laboratory rat, has been domesticated for over 100 years. After so many years of selective breeding, it is likely that the behavior of domesticated rats bears little resemblance to that of their wild counterparts.[17] These concerns about the validity of using the rat as a model for piecing together the evolutionary pathways by which play fighting has been transformed into a complex behavior that serves multiple functions, have to be taken seriously. However, we will show that, with suitable caution, the model that we are proposing can usefully guide us in making play a more comprehensible phenomenon.

Even though rats and other rodents for which sufficient detail on their play is available represent only a fraction of rodent diversity, among these species, play fighting ranges from non-existent to simple to complex. Thus, within a group of related species there exists the range of variation necessary for characterizing the differences between species and these differences can, in turn, be used to infer the steps needed to transform one level of organization to another. However, the paucity of data concerning their natural history and social organization does limit our ability to draw reasonable inferences about the benefits that may have been responsible for facilitating these changes in levels of complexity. Nevertheless, there are broad ways in which these species are known to differ from one another in the wild: some are highly social while others are solitary – this information can provide clues as to the questions that can be asked of lineages of animals where our knowledge of their natural history is greater.

Although few studies have explicitly compared domesticated rats to their wild counterparts, those that have indicate that the behavior of wild and domesticated rats is comparable both in terms of their social affinities and their basic repertoire of behavior patterns. Furthermore, the strains of domesticated rats that are pigmented, rather than fully albino, are most like their wild counterparts, both in their perceptual and motor abilities and in their social behavior.[18] Thus, we can view domesticated rats as stand-ins for their wild counterparts with a reasonable degree of confidence, especially if we are sensitive to any potential differences in behavior among strains of domesticated rats. No doubt some of the details in the patterns of their respective play fighting may differ, but the broad outlines are likely to be similar.

So, despite these limitations, focusing on the play fighting of rats and its comparative context allows us to develop a model to explain how complex patterns of play fighting may have evolved. By characterizing different grades of organization – from simple to complex – we can direct our attention to the brain by asking what changes in brain function would be needed to enable a change in the complexity of play. By taking a brain's eye point of view, we ensure that changes in behavior are grounded in concrete changes in the organ most responsible for that behavior. Furthermore, this approach capitalizes on the wealth of information available, that we have built up over the past three decades or so, on the brain mechanisms that regulate play.[19] The mechanisms responsible for changes in the complexity of play are often valuable clues for those of us seeking explanations as to why such changes may have occurred.

For example, if the brain mechanism responsible for the change in the complexity of play is one that affects many forms of behavioral output, then the output of one of those behaviors may have provided the benefit for the evolution of the change in this brain mechanism, in which case, the change in play may be an incidental by-product. Conversely, if the brain mechanism is specific to play, other aspects of behavioral output are unaffected. In this case, it is likely that it is the benefits accrued from the change in play that are responsible for the evolutionary change in the brain mechanism. A concrete example may best illustrate the point. It would appear that the neural circuitry necessary for humans to evolve language did so to produce spoken language, with written language arising later as a cultural product.[20] However, even though written language is not a direct product of organic evolution, its occurrence depends on the novel use of neural mechanisms that evolved for different purposes. Thus, while spoken language needs to be explained, in part, in terms of the functions for which it evolved, written language can be explained as a by-product of other evolved mechanisms.[21]

Having laid out the basic problem and the approach we plan to use, we can now go on to outline the basic, behavioral properties of play fighting in rats. Armed with this knowledge, we can identify the elements that need explaining and the kinds of explanations that we should seek.

What Do Rats Do When Play Fighting?

Play fighting involves two or more animals seeking to gain some advantage over one another, just as we described for the gorillas at the beginning of this chapter. In figure 1.1, we show an example of play fighting in rats, in which two juvenile males are interacting in a typical fashion. On the left, a rat is seen approaching from the rear (a) and then pouncing towards the nape of its partner's neck (b). However, before contact is made, the defender rotates around the long axis of its body (c) to face its attacker (d). As the attacker moves forward, the defender is pushed onto its side (e), but then rolls over onto its back as the attacker continues to reach for its nape (f–h). From the supine position, the defender launches an attack towards its partner's nape (i), but is blocked by its partner's hind foot (j, k). Following another attempt to gain access to its

Figure 1.1 Two juvenile male rats, at about 35 days old, are shown engaging in a play fight in which they compete for access to the nape of each other's necks. (From Pellis, S. M. & Pellis, V. C. (1987). Play-fighting differs from serious fighting in both target of attack and tactics of fighting in the laboratory rat *Rattus norvegicus. Aggressive Behavior*, **13**, 227–242. © 1987, Wiley; reprinted with permission of Wiley-Liss, Inc. a subsidiary of John Wiley & Sons, Inc.)

partner's nape, the rat on top is pushed off (l) by the supine animal's hind feet (m), which thus enables the original defender to regain its footing (n) and lunge to attack its partner's nape (o). As can be seen, the rats use a variety of offensive and defensive maneuvers as they compete for access to the nape. But what do rats do if they succeed in contacting their partner's nape?

When an attacker succeeds in contacting its partner's nape, it nuzzles this area with the tip of its snout. Occasionally, this nuzzling may be followed by gentle bites (nibbling) or grooming, but in most cases, rubbing the nape with its snout seems to be sufficient to satisfy the attacking rat. In those rare cases when the recipient of such an attack does not defend its nape, the attacking rat will nuzzle the nape, and then leap away. This sequence can be seen in figure 1.2, where the attacker approaches another rat from the side (a), then lunges

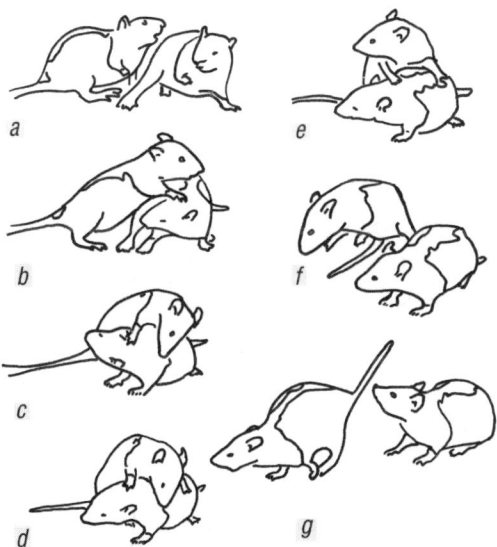

Figure 1.2 Two juvenile male rats, at about 35 days old, are shown engaging in a simple play fight. The animal on the right is shown approaching another rat, then contacting it on the nape and leaping away. The recipient of the attack does not defend itself. (From Pellis, S. M. (1988). Agonistic versus amicable targets of attack and defense: Consequences for the origin, function and descriptive classification of play-fighting. *Aggressive Behavior*, **14**, 85–104. © 1988, Wiley; reprinted with permission of Wiley-Liss, Inc. a subsidiary of John Wiley & Sons, Inc.)

at its nape (b), nuzzles it (c), and leaps away (d). Experimental treatments where the partner is rendered unresponsive to playful attack, but remains active, indicate that nuzzling the nape continues to be actively sought by the untreated partner. Conversely, desensitizing the nape area of a rat with a local anesthetic has demonstrated that in the absence of feeling, the partner's contact on its nape fails to elicit a playful response. Therefore, it seems that a core aim of play fighting in rats is for one animal to contact and nuzzle its partner's nape and for that partner to use various defensive maneuvers to prevent this contact.[22] Moreover, as seen in figure 1.1, another common feature of play fighting, especially in the juvenile phase, is the frequent role reversal between the partners from attacker to defender.[23]

For many animals, the advantage sought during play fighting is to contact one's partner on a specific body target while simultaneously preventing the partner from making that contact.[24] In rats, the target over which the animals contest in play fighting is the nape. With a clearly defined play target, the rats' actions during play fighting can be evaluated in terms of their effectiveness in gaining and blocking access to that target. Of course, we are left with the puzzle as to why nuzzling another's nape is satisfying to a playful rat. To answer this question, we need to understand the play fighting of rats in the context of their social behavior and its development. As play fighting typically involves the use of behavior patterns from other, functional contexts, is the nape nuzzling of rats derived from serious fighting or from some other form of social behavior? In chapter 2, we will characterize the aggressive and the sexual and non-sexual amicable behavior of rats and place playful nape nuzzling in an appropriate context. We will then map the developmental path taken for play fighting and answer questions such as: when does this behavior first emerge and does its content change with age? Do all individuals of a litter follow the same developmental changes in play fighting? In particular, is this behavior expressed in a similar manner in both sexes? Once the basic organization and developmental pattern of play fighting in rats is established, we can then contrast this with other species.

In chapter 3, we will compare the play fighting of rats with that of other rodents. This will achieve several goals. First, it will allow us to identify the common features of play fighting in rodents and those that are unique to rats. Second, by comparing rodents that show varying degrees of complexity in their play fighting, we can identify the kinds of change in neural mechanisms needed to produce the changes in play. Third, by comparing these inferred changes in neural mechanisms with changes in play fighting following selective, experimental damage to various parts of the brain, we can map how the brain actually regulates such play.[25]

Having revealed the organization, development, evolution, and neural controls of play fighting in rats, we will then be in a position to explore the question of the benefits that may arise from engaging in play fighting. In chapter 4, we will examine the delayed benefits of play fighting in the juvenile phase. That is, does play fighting during this phase improve the capacities of the animals when they are adults, and if so, how? In chapter 5, we will consider the possible, immediate benefits of play fighting. That is, how does engaging in play fighting benefit the performer at the time it is playing? If play fighting has immediate benefits, both in the juvenile phase and in adulthood, then it challenges our very conception of play; after all, a key criterion often used to define play is its lack of an immediate purpose. But this very change in perspective directs our attention to how we can differentiate play from its serious counterparts by focusing on the rules by which the behavior is performed, rather than by focusing on its consequences. In chapter 6, we then bring together all the dimensions that have been explored in the previous chapters and use this knowledge about play fighting in rats to develop a model by which we can gain a more general insight into how play may have evolved. Here, we will use several non-rodent examples to illustrate how our rodent model of play fighting provides a useful framework to guide us in studying other examples of play.

In chapter 7, we will explore the implications of what we have learned about play fighting in rats, especially how it relates to humans. For some facets of play fighting, we will argue that what we have learned from rats can be generalized to other animals, including humans. As this is not a crime drama, and keeping the reader in a state of suspense is not our intention, let us outline some of these implications now.

In rats, play fighting is not limited to the juvenile period – although it is most frequent at that age – and seems to be used to achieve both delayed and immediate benefits. By engaging in play fighting, both juveniles and adults can regulate their stress response. That is, by playing, the animals can calm themselves following a frightening, but not too frightening, experience. In addition, male adult rats use play fighting as a means of social assessment and manipulation in the context of dominance relationships. Thus, in at least two ways, play fighting provides the rats with immediate benefits.

During the juvenile period, play fighting is organized in such a manner as to maximize the experience of losing and regaining control over one's own body and that of one's play partner. A lack of such experience leads to a reduction in social competence, as does brain damage to specific brain areas. Conversely, this juvenile experience alters the organization of those same areas of the brain. That is, experience in play fighting in the juvenile period modifies the brain areas involved in social competence. Play fighting thus serves a delayed benefit. A comparison with non-rodent species, especially primates, suggests that both these immediate and delayed benefits may be available to a range of animals, including humans. The evidence for these claims, the details by which rats achieve these benefits, and how these may relate to the neural mechanisms that produce similar benefits in other species, will unfold as you read on. For now, bear in mind that, if people *are* like rats, the current trend in much of the Western world for zero-tolerance of rough-and-tumble play among children is not doing upcoming generations any favors as to the development of their social competence. We need to reconsider such bans, either by loosening them or by identifying alternative ways of gaining the experiences that are typically gained through play fighting. We shall revisit these concerns in chapter 7 once we have a better grasp of what it is that we need to build a socially competent rat.

Finally, in chapter 8, we focus on a re-evaluation of what play is and what it does. Focusing on the arguments as to the delayed benefits gained from playing as juveniles, we conclude that it is not simply a matter of differences of opinion in the preferred skill chosen as the one improved by play experience, but that there are, indeed, fundamentally different ways in which playful experience can affect subsequent behavior. More tellingly, though, we also conclude that no theory to date can account for all the features observed in the play of animals. Just as Hamlet says to his friend, "there are more things in heaven and earth, Horatio, than are dreamt of in your philosophy."[26] Let the games begin!

2

THE PLAYFUL RAT

As we discussed in the previous chapter, play fighting in rats involves attack and defense of the nape of the neck, which, if successfully contacted, is nuzzled with the snout. There are several questions that arise from this behavior, the most obvious being why are rats seemingly excited by nuzzling another rat's nape? And why nuzzle the nape rather than the rump? In order to answer this question about the specificity of the target, we need to cast the behavior of rats onto a larger, comparative framework. For the time being, we will accept that the nape is important and focus on what rats may be deriving from such contact. In addition, we will flesh out the overall pattern of play fighting in rats, such as who plays, when and how.

Tactile contact of another's skin can, in some situations, be pleasant.[1] Let's consider tickling. If someone you know, love, and trust, pins you to the ground and tickles your ribs, you are likely to go into paroxysms of laughter with cries of "no, no, please stop." But, of course, as soon as they do stop, you eagerly anticipate another bout of tickling. In contrast, if a stranger attempted to surprise you by tickling, your response would be very different.[2] The response to tickling is similar in rats. Whether nuzzled by another rat or tickled by a familiar experimenter, rats appear to enjoy the experience – they will briefly tolerate the contact and run away but then come back for more. Furthermore, when undergoing such an experience, they emit an ultrasonic call akin to human laughter and their brains release neurochemicals associated with pleasure.[3] So, it seems that being playfully nuzzled is an enjoyable experience and must be the engine that drives rats to want to play. But what does the rat doing the nuzzling gain?

Play fighting in rats involves one animal (i.e. attacker) trying to contact the other's nape, while the recipient (i.e. defender) uses various maneuvers to avoid this contact. The roles of attacker and defender can alternate not only between bouts of play fighting, but also within a bout (see figure 1.1). Most of the time (80% or more), the recipient defends itself against nape contact, but occasionally, it does not, and in such situations, the attacker is free to contact its nape briefly (figure 1.2). Clearly, nuzzling and avoiding being nuzzled are what organize play fighting, and, as we discussed above, being nuzzled is pleasurable. But this means that nuzzling the nape of one's partner must also be enjoyable, given the great lengths to which rats will go to do so. Indeed, if juvenile rats are socially isolated for a few hours or days, and thus deprived of the opportunity to play, they will exhibit elevated levels of play fighting when they are re-housed together.[4] Moreover, if play-deprived rats are re-housed with a partner that has been chemically treated so that it is active, but non-playful, the untreated animal will still run up to its partner and contact its nape at a heightened frequency.[5] Even without a playful response by the partner, gaining access to the nape and nuzzling it seems to be rewarding in itself. If we return to our tickling example, tickling someone else seems to be just as pleasurable as being tickled. So, both giving and receiving nuzzling likely taps into the same pleasure system that is engaged by tactile contact.[6]

But is giving and receiving nape contact all that is needed to account for the genesis and maintenance of play fighting in rats? It seems not. After prolonged exposure to a drugged rat that does not resist having its nape contacted, an attacker finds the experience disconcerting. It will even begin to dig at the substrate, which is an anti-predator behavior often used by rats in anxiety-inducing situations.[7] Thus, the lack of a suitable response from its playmate appears to distress a rat. Therefore, while nuzzling the nape of another rat is rewarding, some active resistance to such contact by the recipient of that nuzzling makes it even more so. Quite possibly, gaining access to one's partner's nape, but doing so while that partner is actively resisting, makes playful attack more exciting. The rat's excitement is likely further enhanced if there is the chance that not only will its attack fail, but also that its partner will launch its own attack. Thus, a play fight involves attack, defense, and counterattack.

Engaging in playful attacks with all these options must be part of a rat's expectations about the world of play, and a rat must experience at least some fully developed play fights in order to activate its pleasure system completely when it is triggered by playful tactile contact.

We should bear in mind that play fighting is a combination of competition and cooperation. Recall your own childhood and the play partners over whom you easily gained the advantage and those with whom you never gained the advantage – both these individuals were likely the ones whom you did not find exciting as play mates. In contrast, those individuals with whom gaining the advantage was a challenge, but who did not make it impossible for you to win, were likely the ones with whom you played most often. In the case of rats, in order for play fighting to be fully rewarding, not only is contact with the nape necessary, but there must also be some degree of competition for access to the partner's nape.

The options of how to avoid nape contact are not unlimited; rather, they involve a distinctive set of tactics. These vary in their usage in predictable ways during development, between the sexes, and, in some situations, across individuals.[8] Although playful attacks to the nape may be directed from any orientation to the partner's body, including directly from the front or from the rear, attackers most commonly approach from the side.

When the attacker approaches its partner's nape from a perpendicular angle then, just before contact is made, the partner has the full freedom to execute all available defensive tactics. It is from this configuration that the alternative options available to the defender are therefore most evident to an observer.[9] So, what are the tactical options available? The simplest thing to do, of course, is nothing. Go on your merry way and ignore your partner's attack. Although this does happen, in 80% or more of attacks, the recipient performs some defensive action. Because the target of attack is the nape, the recipient performs defensive maneuvers that block access to this part of its body. There are generally two ways of doing this. First, the defender can move its nape away from the attacker while simultaneously withdrawing its head and upper body. This strategy can involve several tactics: the defender can run or leap away, but more commonly, it can swerve away to the side (figure 2.1A). In this case, it is moving both its nape and its head away from its attacker – we refer to this as "evasive defense." Second, the defender can move its nape away while simultaneously turning its head so as to face its attacker – we refer to this as "facing defense." This second style of defense involves two types of tactics.

Figure 2.1 Three types of defense to a playful nape attack are shown in 61-day-old male rats (A–C). A. Evasion. Following the attacker's lunge at the nape (a, b), the defender swerves away from the attacker (c). B. Complete rotation. A nape contact from behind (a) leads the defender to rotate (b, c) until lying supine and blocking the attacker with its outstretched paws (d). C. Partial rotation. A lunge to the nape from the side (a, b) is followed by a rotation of the head, neck and shoulders by the defender, withdrawing the nape from the attacker's snout (c). (From Pellis, S. M., Pellis, V. C., & Whishaw, I. Q. (1992). The role of the cortex in play fighting by rats: Developmental and evolutionary implications. *Brain, Behavior & Evolution*, **39**, 270–284. © 1992, Karger, reprinted with permission of S. Karger AG, Basel.)

The first tactic involves the defender performing a rotation around the long axis of its body, beginning at its head and then progressing down the entire length of its body, while facing the attacker. This can take one of two forms. The rotation can be completed so that the defender ends up lying on its back with all four paws in the air (figure 2.1B). From this position, it can use all four limbs to hold its partner at bay. Alternatively, the defender can only partially rotate, keeping one or both of its hind feet on the ground (figure 2.1C). From this position, it can then push its partner with either its forepaws or its hip, or rear onto its hind legs to face the attacker. The second tactic involves the defender swerving towards its approaching partner while continuing to stand on all four paws or on its hind paws. From this position, the defender can block its partner's movements with its forepaws (figure 2.2).[10]

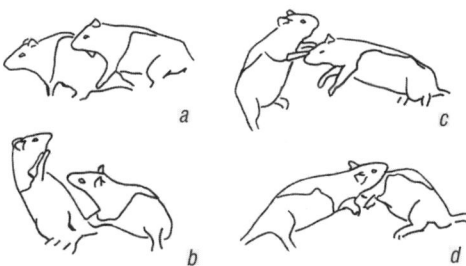

Figure 2.2 This is a form of facing defense in which the defender turns to face its attacker while pivoting on its hind legs (a, b). From this position, the defender can block access to its nape (c) and launch a counterattack (d). (From Pellis, S. M., Pellis, V. C., & McKenna, M. M. (1994). A feminine dimension in the play fighting of rats (*Rattus norvegicus*) and its defeminization neonatally by androgens. *Journal of Comparative Psychology*, **108**, 68–73. © 1994, APA, adapted with permission.)

Note that when the rat swerves towards its attacker, or when it partially rotates its body, its defensive action can result in an upright stance. Similarly, a rat adopting either of these tactics can be bowled over by the attacker, which can lead the defender to lie on its back. This action mimics the end result of the complete rotation. Therefore, when scoring the tactics used, it is important to identify the tactic by referring to the actual movements performed by the defender, and not by the outcomes of the combined movements of the two participants. For example, when a juvenile rat engages in play with a young adult rat, the superior body weight and size of the partner usually results in the juvenile being pushed over onto its back, irrespective of the defensive tactic adopted by the juvenile. By defining the tactics used by their constituent movements, it is possible to score the preferences of the performer, rather than the outcomes that arise from factors beyond the performer's control.[11] Armed with this basic toolkit of movement patterns and a caveat for how best to evaluate those that are used by rats, we can begin to examine the various factors that influence the frequency and content of play fighting.

The Development of Play Fighting

Play fighting in rats follows an inverted U shape, appearing in the animals' repertoire around seventeen to nineteen days after birth (as a reference point, the eyes open around twelve days), and gradually increases in frequency until reaching its peak between thirty to forty days, when they play for up to an hour a day. The frequency of play then decreases again, so that by around puberty (sixty days), they only play for a few minutes a day. While both males and females follow this developmental sequence, most studies have found that the absolute frequency of play is higher in males, especially in the juvenile period.[12] This pattern of change in the frequency of play fighting during development can be broken down further by examining the frequency of launching attacks and the likelihood of defending against attacks. This reveals that the changing frequency of play fighting mostly arises from changes in the frequency of attacks, not by changes in defense, which remains relatively stable (see figure 2.3). Nevertheless, while the likelihood of defense does not change with age, there *are* changes in the likely use of some defensive tactics.

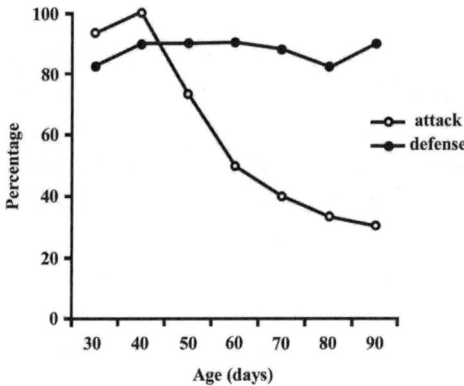

Figure 2.3 The age-related time course of playful attack and playful defense is shown for male rats. Although the probability of defense remains constant, the incidence of playful attack peaks between 4–5 weeks after birth (30–40 days). Note that while the y-axis represents the percentage of both attack and defense, the values given represent something different for attack and defense. For playful attacks, the percentage is shown at each age as a fraction of the total at all ages – that is, the rats launched more attacks at 30 and 40 days than they did later in development. In contrast, for playful defense, the values represent the percentage of playful attacks that elicit defense as a proportion of the total number of attacks at that age. (From Pellis, S. M. & Pellis, V. C. (1991). Attack and defense during play fighting appear to be motivationally independent behaviors in muroid rodents. *The Psychological Record*, **41**, 175–184. © 1991, Southern Illinois University, adapted with permission.)

At all ages, evasion accounts for 20–30% of defensive responses, while swerving towards the opponent is used in about 5–10% of cases. For both males and females, when play fighting first emerges around weaning, the partial rotation is the more frequently used tactic, but then, during the juvenile period, complete rotation is the more frequently used tactic. With the onset of puberty, males switch back to using the partial rotation most frequently, whereas, in females, it continues to be used less often than the complete rotation tactic (figure 2.4).[13]

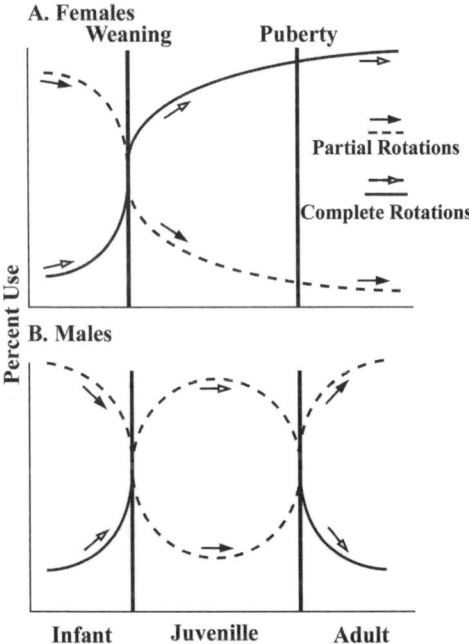

Figure 2.4 Age-related changes in tactics of defense show that for both males and females, the partial rotation tactic is the most frequent around weaning, but the likelihood of its use as the predominant type of facing defense in the juvenile period is then reduced as the complete rotation tactic becomes the more frequent defense. Following puberty, the sexes differ: females continue to use the complete rotation tactic more often (A), but males switch to using the partial rotation more often (B). (From Pellis, S. M. (2002). Sex-differences in play fighting revisited: Traditional and non-traditional mechanisms for sexual differentiation in rats. *Archives of Sexual Behavior*, **31**, 11–20, Figure 1. © 2002, Plenum Publishing Corporation, copied with kind permission of Springer Science and Business Media.)

Was it Good for You, Big Boy?

So, let's return to our question: why is nuzzling the nape rewarding? To deal with this, we first need to consider what the label "play fighting" really means. The term "fighting" conjures up visions of a bar brawl, a physical situation in which one or more persons attempt to induce injuries in others. The addition of the term "play" in front of "fighting" suggests a situation in which the individuals are exhibiting behavior which resembles fighting, but do not seem serious in their intentions. Viewed in this way, we could assume that, when engaged in play fighting, animals are essentially mimicking what they do in serious fighting. If we consider dogs, cats, and most other species with which readers are familiar, this is a reasonable assumption; for many species, the body targets competed over during serious fighting and play fighting are the same.[14] Rats, however, appear to be an exception to this general pattern.[15]

When rats engage in serious fighting, the primary targets of attack are the face or the flanks and lower back, which, if contacted, are bitten. Similarly, a defending animal may retaliate with bites directed at the side of its attacker's face.[16] Thus, neither the offensive nor the defensive targets present in serious fighting are contested during play fighting. Rather, both the attacking and the defending rats focus their attention on each other's napes.[17] Play fighting in rats does not, therefore, mimic serious fighting. Where does this preference for the nape come from then? There are two possible social situations from which playfully contacting the nape may be derived. In adult interactions, the nape is contacted during social grooming – especially when an unfamiliar animal is placed into the home cage of another. After sniffing the intruder to determine that the rat is unfamiliar, the resident may begin to groom the newcomer. As grooming progresses towards the nape, it becomes rougher; what begins as nibbling of the newcomer's fur turns into tugs at the underlying skin. This has justifiably been termed "aggressive grooming." Close inspection of nape contact during play fighting reveals that this nibbling and tugging does occur occasionally, but that the majority of the contact involves one rat nuzzling its snout into the fur of its partner's neck.[18]

The other possibility is that nape contact during play is related to sexual behavior. During sexual encounters, males will approach females and sniff them. In these circumstances, sniffing serves both to identify the female and determine her state of sexual readiness. If she is ready to mate (that is, is in estrus), the male will then switch to nuzzling the female's nape as a prelude to mounting her. But the female rat is not just a passive recipient of the male's overtures – rather, she uses a variety of behaviors, such as ear wiggling and darting in front of the male, to solicit the attention of the male of her choice. Then, by resisting his overtures and leading him on a merry chase, she can pace the sequence of mounting and copulation. For our purposes, what is important is that it is in sexual encounters that we see nuzzling akin to that present in play fighting. That play fighting in rats may be mimicking sexual behavior is reinforced by the observation that when females defend themselves against nape contact, they use the same range of defensive tactics used in play fighting. To be fair, the same tactics are also seen in aggressive encounters,[19] so, to make the case for the similarity between play and sex, we must examine the issue more closely.

In serious fighting, it is not easy for one rat to press an attack to bite an opponent's rump successfully. The dynamics of serious fighting are best understood in the context of a resident male warding off an intruding male that is entering his territory. In this situation, the resident does all the attacking and the intruder does all the defending. In order to block access to its rump and lower flanks, the defending rat can adopt an upright position, in which, by standing on its hind feet, it can readily pivot around and so keep its teeth continually aimed at the side of its attacker's head. As the resident lunges at the defender's lower flank, the defender can swoop down and deliver a retaliatory bite to the side of the resident's face. Thus, the problem for the attacker is how to deliver a bite without itself being bitten. One approach is to sidle up to one's opponent while maintaining a broadside or lateral orientation (figure 2.5). From this position, the attacker can then press its flank into the opponent's ventrum, and, if its opponent is pushed off balance, the attacker can rapidly take the opportunity to deliver a bite to the exposed flank. Because in this resident–intruder paradigm the defender does not bite its attacker's flanks and rump, it is safe to press these parts of its body into the defender. This strategy has two beneficial effects for the attacker. First, in employing this strategy, the side of its own head is kept well away from the defender and second, by using its body to upset the defender's balance, the attacker ensures that there is a brief moment in which the defender's flank is exposed and the defender is in a weak position from which to retaliate. Not

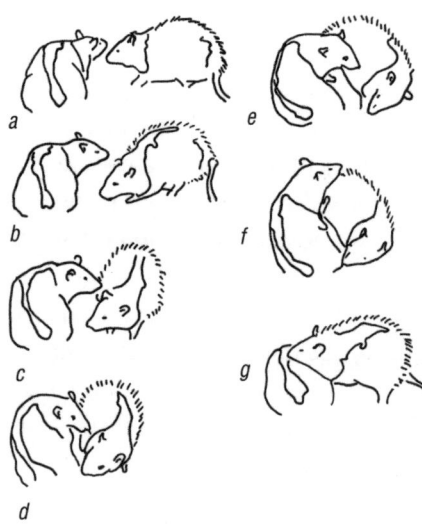

Figure 2.5 In serious fighting in adult rats, an attacker often approaches a defender with a lateral orientation (a–d). From this orientation, the attacker can avoid being bitten on the face (e), can reach around to the defender's lower flank (f) and can lunge to deliver a bite (g). Note that the attacker's fur is raised (i.e. piloerected) indicating its aggressiveness towards the opponent. (From Pellis, S. M. & Pellis, V. C. (1987). Play-fighting differs from serious fighting in both target of attack and tactics of fighting in the laboratory rat *Rattus norvegicus*. *Aggressive Behavior*, **13**, 227–242. © 1987, Wiley; reprinted with permission of Wiley-Liss, Inc. a subsidiary of John Wiley & Sons, Inc.)

surprisingly, in such encounters, defensive upright positions and offensive lateral orientations are common.[20]

Encounters can become more complex when both animals are on the offensive, such as when competing for a desired resource on neutral ground. Each animal has to organize its actions so that it can simultaneously avoid bites to its own rump and yet be able to deliver bites to its opponent's rump. Further, each animal has to deal with the uncertainty that the opponent may switch from attempting to access its rump to lunging, frontally, at its face. Thus, both animals need to determine how to reduce the risk of retaliatory bites to the sides of their face, while attempting to bite their opponent's rump. Even with these extra complications, upright defensive postures and lateral orientations feature prominently in serious fighting.[21] This is not true for play fighting – when such postures are used, they are greatly modified by the fact that attack and defense revolves around the nape. Indeed, the tactics that are used for defense in play fighting are more similar to those used by female rats during sexual encounters so as to avoid nape contact, and both only superficially resemble these tactics when used in serious fighting.[22] Thus, nuzzling the nape during play fighting mimics the competition for nape contact in sexual encounters. The comparative evidence that further supports this conclusion will be examined in the next chapter, and will reveal that the play fighting of rats is not aberrant.

Rethinking the Play Fighting of Rats

In rats, play fighting involves attack and defense of a body target that is competed over during adult sexual interactions. A potential terminological quagmire lies ahead, because we need to consider whether such play in rats should still be termed play fighting. Because play fighting has typically been viewed as the mimicking of serious fighting, it has also been termed "aggressive play." When mounting resembling that seen in copulation is engaged in playfully, it has been termed "sexual play." Mounting during play in rats is relatively rare, and is seemingly distinguishable from play fighting. In species such as dogs where the targets of playful attack are the same as those present in serious attack, this can be thought of as aggressive play, whereas competing for mounting can be thought of as sexual play. The two sets of actions clearly mimic two distinct, functional sets of adult behavior. This is also true for some rodents, such as ground squirrels, whose actions during play fighting alternate between directing bites to the shoulders, as adults do during serious fighting, and attempting to mount one another, as adults do during sexual encounters.[23] Rats differ in that during play fighting, they attack and defend a body target that is contacted during adult sexual encounters, but unlike mounting, which represents the beginning of the culmination of the sexual act, the nape is a body area that is contacted in the precopulatory phase of the sexual encounter. Thus, while it is not obviously sexual, their play fighting only superficially resembles the aggressive play of species such as ground squirrels. Yet, since play fighting in rats involves mimicking the early stages of the sexual encounter, it is more appropriately labeled sexual play.

Although logical, a simple label change can in itself be misleading. No matter whether the targets involved in play fighting are derived from aggressive or sexual sources, it remains a competition to gain some advantage. For some species, the play targets competed over are the same as those which are typically used when attacking and killing prey.[24] Irrespective of the advantage sought, when animals engage in play fighting, they are constrained to follow certain rules that ensure that the play fighting remains playful.[25] Therefore, we propose that when the features common to competitive play are being emphasized, it is appropriate to continue to use the term 'play fighting,' no matter whether the advantage sought involves a body target derived from aggression, sex, or, for that matter, predation. However, when the idiosyncratic properties of play derived from sex, aggression, or predation are being emphasized or compared, then it is appropriate to refer to them, respectively, as sexual play, aggressive play, or predatory play. It is important to note that this schema is based on the intrinsic content of the play, not on its functional consequences. For example, in some species, play fighting may be used for courtship purposes. However, in some species, courtship play is derived from patterns of sexual behavior, whereas in others, it is derived from aggression. Thus, terming this behavior "sexual (or courtship) play", because it is used in a sexual context, confounds the potentially differing origins of the patterns of behavior used in that play. In such a situation, when using the term sexual play, it becomes unclear whether an author is referring to the organization or the function of the play. For these reasons, we are opting to retain the flexibility of using the term play fighting irrespective of the origin of the content of the play, be it sexual, aggressive, predatory, or other. As the need arises, the terms sexual play and aggressive play will be used, but, in all cases, when these terms are used, they will reflect the *content*, not the *function*.

Sex Differences in Play Fighting

Having clarified the terminological issues, we can now re-examine the basic features of play fighting in rats. First, even though play fighting mimics precopulatory behavior, which involves the male contacting the nape of the female and the female avoiding such contact, in play fighting both males and females attack and defend the nape. This is not as peculiar as it may seem. Many behavioral differences between the sexes are like this: both sexes are capable of expressing the behavior, but one sex is more likely to do so than the other. Even behaviors that are most associated with one sex follow this pattern. In rats, during sex, males mount females and females adopt a posture called lordosis, in which the back is arched downwards, raising the rump upwards and exposing the vagina. Yet, sometimes, females mount males and males perform lordosis, making these behaviors sex-typical, rather than sex-exclusive.[26] Recall that for rats, as most studies report that males engage in more play fighting than do females, play fighting may be considered a male-typical behavior.[27]

Second, while in many species play fighting is most frequently engaged in during the juvenile period, in rats it persists into adulthood. Because play is most commonly associated with the juvenile period, it has been traditionally thought of as the property of a young animal. Analogous to sex differences, where certain behaviors may be thought of as male or female-typical, play may be thought of as juvenile-typical. Nonetheless, play, and in particular, play fighting, occurs in the adults of many species, and in some rare cases, play fighting is actually more common in some species among adults! Much of the play fighting seen in the adults of these species involves an interaction with an immature individual, but in other species, it is preferentially directed towards other adults.[28] Because rats retain play fighting into adulthood, the content, causes, and functions of this behavior can be examined in both the juvenile and post-juvenile period of life. The continuance of play fighting into adulthood will figure greatly in our assessment of the functions of play fighting in rats, but this story will be dealt with in the chapters to come. For now, we will explore some further details about sex differences in the play fighting of rats.

The traditional model that accounts for sex-typical anatomy, physiology, or behavior in mammals is that all embryos begin as females. That is, in this model, unless the developmental process is instructed otherwise, the female is produced as the default condition. However, when the correct genetic cue is given, a cascade of events can transform the developing organism into a male. This genetic trigger induces the gonadal tissue to develop into testes rather than ovaries. The testes then produce large quantities of male-typical sex hormones,[29] and this, in turn, leads to the organization of sex-specific neural circuits. Thus, at the correct age, males will be more likely to perform male-typical behaviors and less likely to perform female-typical behaviors. To continue the example of mounting behavior discussed above, these changes make males more likely to mount and less likely to perform lordosis when they themselves are mounted. These organizational effects on the nervous system, and, subsequently, on the behavior, occur before the animals are capable of performing these actions, when the young are undergoing fetal development. Later, the same hormones may trigger these neural circuits to emit their differential behavioral expressions.[30]

Either by suppressing fetal testosterone production in the last few days before birth or by castrating male infant rats in the first few days following birth, males will develop a female-typical pattern of behavior. Not only will these males be unlikely to mount receptive females and likely to show lordosis when they themselves are mounted, but they will also display the female-typical pattern and frequency of play fighting. Thus, the male-typical pattern of play fighting is dependent on the early developmental organizational effects of testosterone. Similarly, augmenting the amount of testosterone exposure in newly born female rats will lead to the development of females with a male-typical pattern in the frequency of their play fighting. Studies that have examined the effects of these hormone manipulations on the two features of play fighting – attack and defense – have found that it is the frequency of playful attack that is most sensitive to these hormonal influences. Testosterone reduction in males leads to a female-typical frequency of playful attack, and testosterone augmentation in females leads to a male-typical frequency. In females, removing the ovaries does not derail the females from developing the female-typical frequency of playful attack. Furthermore, neither testicular nor ovarian manipulations affect the likelihood of defense when playfully attacked.[31] Thus, sex differences in the frequency of play fighting follow the traditional model for sexual differentiation. However, for age-related changes in the tactics of defense, the story is more complex.

The age-related changes that we discussed earlier follow the same pattern in both males and females until puberty. When play fighting first emerges at around the time of weaning, both males and females are more likely to use the partial rotation tactic. By the peak juvenile period, both are more likely to use the complete rotation tactic. With the onset of puberty, the pattern is once again reversed in males, while females retain the juvenile pattern of play (see figure 2.4). Castration of males shortly after birth suppresses the pubertal change in defensive tactics, causing them to retain the juvenile pattern, like females do, into adulthood. However, if the castration occurs shortly after weaning, there are no changes to the normal development of the male-typical pattern of play. That is, testicular hormones masculinize the brain early in development, but once it is so organized, they are not necessary to activate the changes in the pattern of play. These effects are consistent with the way in which testosterone is necessary to masculinize the frequency of play, and with the traditional model of how sexual differentiation occurs.[32] But this is not true for females.

If females are augmented with testosterone shortly after birth, the female-typical pattern of retaining the juvenile-typical defense following puberty is not affected. However, if their ovaries are removed either shortly after birth or following weaning, this pattern is affected. These females switch to the male-typical pattern, so that following puberty, they will perform more partial rotations than complete rotations. Therefore, for females to retain their female-typical pattern of defense, there has to be an active suppression by ovarian hormones. This is inconsistent with the traditional model of sexual differentiation.[33] These findings support a growing body of evidence that demonstrate that the female mammal is not the default condition for all traits.[34]

More importantly, for our current story, these findings reinforce the view that play fighting is a composite of distinct behaviors with distinct processes regulating them. Such a mosaic conceptualization of play fighting will become useful in later chapters when we explore how rat-like play fighting may have evolved. Before we switch to constructing the comparative framework necessary for this endeavor, we first need to consider whether critical neural machinery might be needed in order to play like a rat. To gain insight on this matter, it is useful to compare the play fighting seen in rats with the more rudimentary form of such play present in the house mouse.

Turning a Mouse into a Rat

The rat and the mouse are the two most commonly studied laboratory rodents. In fact, with the mapping of the mouse genome complete, and with the ability to add, delete, and manipulate the activation of specific genes, the mouse has become more popular than ever. What a wonderful model the mouse would be to study the genetic–neural–behavioral processes that generate play fighting! Unfortunately, the mouse has a relatively impoverished social play repertoire. Instead, mice are renowned for their solitary locomotor play. Indeed, this behavior is so common following weaning, that it has been

referred to as the "popcorn" phase – an evocative term, which readily conjures up images of little, white bundles hopping, jumping, and running.
Closer study of this behavior has revealed that this bopping about is not
completely asocial. Rather, much of the jumping is contagious and the running is synchronized, which suggests that there is some social coordination to
all this play.[35] However, this is still a far cry from the complex wrestling that we
have described in the play fighting of rats.

In a small minority of cases (2%), mice do not run simultaneously in the
same or different directions to the others; rather, one mouse runs towards
another, aiming its snout at the other's neck. As this "attacker" approaches,
the "defender" swerves and runs away, but then may run towards the original
attacker, which will, in turn, swerve and run away. Under some conditions, a
mouse may leap onto another from a raised platform, which also leads to fleeing and chasing. So, even though infrequent, mice have the basic rudiments of
play fighting present in rats – attacks directed at the neck and evasive
defense.[36] There are two possible explanations for the impoverished play
fighting of mice.

One possibility is that during such playful encounters, mice may simply
not possess the behavioral repertoire necessary to do more than rudimentary
approaches and withdrawals. An alternative explanation may be that a mouse
has a richer repertoire of responses, but due to the limitation in its partner's
playful attacks, does not have the opportunity to express the behavior that is
latent. This latter possibility is not as peculiar as it may seem at first reading.
Female rats tend to respond to the approach of an attacking play partner
sooner than males. As a consequence, females are more likely to use evasive
tactics effectively and employ the swerving toward, rotation tactic of facing
defense. Because males wait longer to react, they essentially begin their
defense when their partner has already made contact and so are necessarily
more limited in their options. These perceptual differences between the sexes
partly account for the sex differences in play fighting behavior.[37] Similarly, it
is possible that mice begin to respond early to their attacker's approach, and
so are rarely, if ever, constrained to use facing defense, and this is what leads
to the absence of playful wrestling. But what if the mice could be placed in a
situation in which they had no choice but to extricate themselves from an
actual nape contact? The crucial experiment to test this possibility has been
done.

Experimenters placed young mice into litters of rat pups that were being raised by rat mothers. Given that they had both mice and rat partners with which to play, these young mice had the opportunity to express all the behavior of which they were capable, since their rat "siblings" attacked them in the usual manner for rats. In these conditions, whether approached by mice or rat attackers, the mice did not use the facing defense to ward off their attackers, not even when their rat foster-siblings actually jumped on top of them and nuzzled their napes. So, even when given the opportunity, mice do not engage in rat-like patterns of defense and do not exhibit rat-like play fighting.[38] Therefore, it does seem that mice do not play like rats – not because they do not have the opportunity to do so, but because they lack the necessary behavioral repertoire. It seems highly likely that this more impoverished behavioral repertoire arises from the absence of the necessary neural machinery. If this were true, we would ask the question differently: what neural mechanisms do we need to add to the brain of a mouse in order for it to play like a rat? To deal with this question effectively, our comparisons need to take into account a range of rodent species with more gradations of complexity in their play fighting.[39]

3

TURNING A MOUSE BRAIN
INTO A RAT BRAIN

At the end of the preceding chapter, we found that even when provided with the opportunity, a mouse cannot play like a rat. This suggests that novel control mechanisms have been added to the rat brain so that rat-typical play fighting is possible.[1] If we cast our comparative net more widely, we can characterize the components that are missing in the mouse, but present in the rat. Using this approach, we can also determine whether one simple change is necessary to go from one rodent to the other, or whether multiple steps are needed. Either way, a comparative analysis can identify what the necessary neural mechanisms are for such a transition. The first thing that we notice when looking at more species of rodents is that the mouse does not have the simplest play fighting. Some species show no play fighting whatsoever, not even the rudimentary pattern that is seen in mice. So, our journey will not be from rudimentary mouse play to complex rat play, but from no play at all to complex rat play, with the mouse representing some intermediate state. Such an insight makes our journey simpler, because it alerts us to the prospect that building a brain capable of generating complex, rat-like play fighting involves more than one change in the brain (although all the necessary changes could be produced in one fell swoop). What are those changes and how can they be identified and characterized?

By comparing the play fighting of sixteen murid rodents, we revealed that some do not have any play fighting whatsoever, and among the remaining

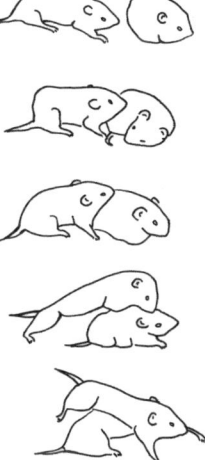

Figure 3.1 Playful attack in European voles, involving "nosing" of the nape area without any associated defense is shown. The attacker approaches from the rear, noses the partner's nape and then leaps away. Note the similarity to the sequence depicted in Figure 1.2 for rats. (From Wilson, S. (1973). The development of social behaviour in the vole (*Microtus agrestis*). *Zoological Journal of the Linnean Society*, **52**, 45–62. © 1973, adapted with permission from Blackwell.)

ones (twelve out of the sixteen), there are gradations of complexity. To characterize this complexity, the various components of play fighting in rats that we described in chapter 2 can be mapped onto the other species.[2] At the simplest level, a species may have attack, but no defense (figure 3.1). Added complexity arises from the recipient of an attack defending itself, but this can also vary in complexity – the defender may use only evasive defense (e.g. house mouse), which leads to play fights that have a pattern of approach and withdrawal that appears simple. To this, may be added facing defense; this leads to wrestling and so more prolonged playful contests involving lots of bodily contact. Finally, the defender may not only use facing defense to protect its play target, but once having successfully done so, it may launch a counterattack at its partner's play target (figure 3.2). Rats can engage in all four levels of interaction, from simple play (attack only, see figure 1.2), to varying degrees of rudimentary play (attack and evade or attack and facing defense, see figures 2.1 and 2.2), and, finally, to complex play (attack, facing defense and counterattack, see figure 1.1). As shown by the deer mice in figure 3.2, the rat is not the only rodent with a complex pattern of play.

Closer inspection of play fighting among murid species reveals that they differ qualitatively (e.g. those species that defend versus those that do not), and also quantitatively (e.g. some species are more likely to defend more often when attacked than others). This is well illustrated by comparing rats with

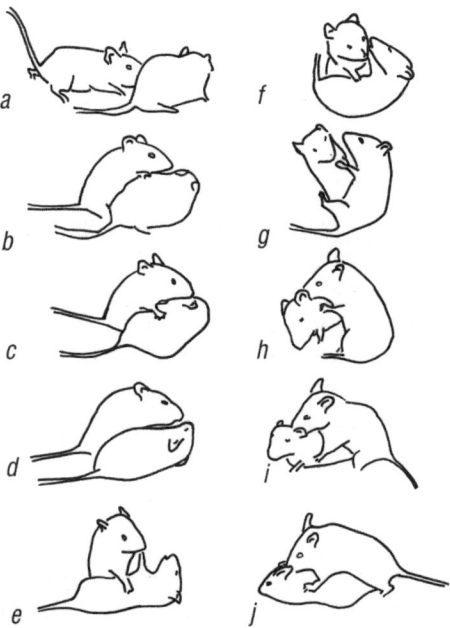

Figure 3.2 This sequence of play fighting in deer mice shows that, like rats, their play fighting involves all the elements that are needed for complex play: attack (a–c), defense (d, e) and counterattack (f–h), which can lead to a role reversal (i, j). (From Pellis, S. M., Pellis, V. C., & Dewsbury, D. A. (1989). Different levels of complexity in the playfighting by muroid rodents appear to result from different levels of intensity of attack and defense. *Aggressive Behavior*, **15**, 297–310. © 1989, Wiley; adapted with permission of Wiley-Liss, Inc. a subsidiary of John Wiley & Sons, Inc.)

deer mice. Both these species have all three components of play fighting (attack, defense, and counterattack), but when their respective play fighting is examined under comparable conditions, so that the differences between them do not arise from differences in opportunities, rats engage in the most complex sequences more often (figure 3.3). Applying this logic of comparison (qualitative and quantitative) to all the murid rodents for which data are available reveals the gradations in complexity of play fighting across species (table 3.1). Here, species can be ranked in ascending order of the complexity of their play fighting, and it is clear that to go from species that do not engage in play fighting (e.g. some hopping mice) to species with the most complex play fighting (e.g. rats, Syrian golden hamsters), involves a graded series of

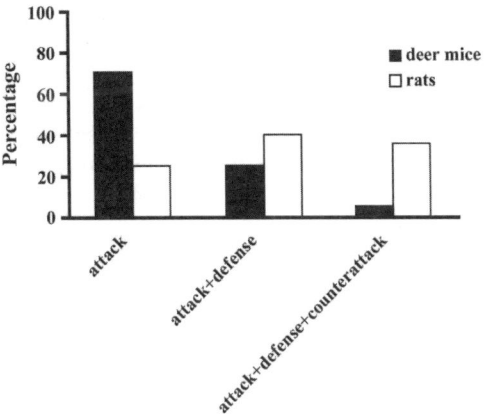

Figure 3.3 These bar graphs show the percentage of play fights that involve attack only, attack and defense, and finally, attack, defense and counterattack. Even though both rats and deer mice engage in play fights at all levels of complexity, rats engage in the complex patterns more often. (From Pellis, S. M. & Pellis, V. C. (1998). The play fighting of rats in comparative perspective: A schema for neurobehavioral analyses. *Neuroscience & Biobehavioral Reviews*, **23**, 87–101. © 1998, adapted with permission from Elsevier.)

additions in novel control mechanisms. The importance of adding such control mechanisms is illustrated by two other differences in the complexity of play fighting in these species, but before examining these differences, we need to place play fighting in rodents in a broader context.

Before proceeding on to comparisons of play fighting among these species, it is important to note that while we would like to take credit for having thoughtfully chosen species on some systematic basis, alas, we cannot. As discussed in chapter 1, the rodent order of mammals contains many species, with the murid rodents being the subgroup containing the most. Unfortunately, the murid rodents also tend to be the smallest members of the order, the ones most likely to be active after sunset, and so the group least studied in the wild. The main criterion we used to decide on which species to compare was their accessibility. Nonetheless, we did try to make sure that we had species representing at least five different families of murid rodents, which provided a range of species with diverse evolutionary histories, in some cases spanning at least forty million years. What was most important to us was that we had species that ranged from no play fighting to complex play fighting and much gradation in between. This diversity provides a framework

Table 3.1

Comparison of 12 species of murid rodents that engage in playful attack. Not only does such playful attack vary in frequency between species, but the species also differ in the frequency with which they use different types of defensive tactics. Indeed, some do not defend at all. (From Pellis, S. M., & Pellis, V. C. (1998). The play fighting of rats in comparative perspective: A schema for neurobehavioral analyses. *Neuroscience & Biobehavioral Reviews*, **23**, 87—101. © 1998, adapted with permission from Elsevier).

Species	Playful attack	Evasive defense	Supine defense
Laboratory rat (*Rattus norvegicus*)	++++	+	++(+)
House mouse (*Mus musculus/domesticus*)	(+)	++++	None
Jumping mouse (*Notomys alexis*)	++	None	None
Mongolian gerbil (*Meriones unguiculatus*)	++	++	(+)
Fat sand jird (*Psammomys obesus*)	++	++	+
Grasshopper mouse (*Onychomys leucogaster*)	+++	++	+
Deer mouse (*Peromyscus maniculatus*)	+++	+(+)	+(+)
Syrian golden hamster (*Mesocricetus auratus*)	+++	(+)	+++(+)
Djungarian hamster (*Phodopus campbelli*)	++ (+)	++	+
European field vole (*Microtus agrestis*)	++	None	None
Montane vole (*Microtus montanus*)	+++	(+)	+
Prairie vole (*Microtus ochrogaster*)	+++	+(+)	+(+)

Legend: + represents one arbitrary unit and (+) represents a half a unit. Although arbitrary, the distribution of units allows comparison both within and between species.

with which to try to piece together how it is possible to evolve a playful brain, a matter dealt with at length in chapter 6.

The Peculiarities of Play Fighting in Murid Rodents

Although play fighting conveys the impression that it should mimic serious fighting, we showed in chapter 2 that this is not true for rats. Instead of targeting those areas attacked in serious fighting, playing rats target the nape, which mimics adult sexual encounters.[3] For rats, and all other murid rodents that have been studied, serious fighting involves biting attacks directed at the flanks and the rump of the body, and, if the attacker is highly agitated, it may even direct bites at its opponent's face.[4] Like rats, all the other murid rodents for which play fighting has been described mostly contact the adult-typical targets used by males during sexual encounters.[5] Furthermore, whereas some species are like rats in that they contact the nape (e.g. voles, deer mice), other species compete over contact with different targets. For example, Syrian golden hamsters nibble the cheeks, fat sand jirds (a type of gerbil) make gentle snout-to-cheek contact, and grasshopper mice nuzzle and groom the shoulders, neck, and the side of the head. The most compelling of all is the "kissing" of the Djungarian hamster. These rodents compete to lick their partner's mouth during both playful and sexual encounters (figure 3.4). Despite the variation in their targets, what binds all these species together

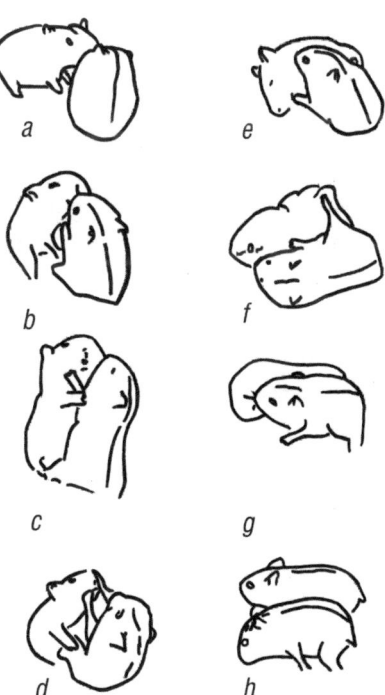

Figure 3.4 A sequence in a pair of juvenile Djungarian hamsters shows that their play fighting involves competition for access to the mouth. The attacker approaches from the front and reaches to contact the partner's mouth orally (a, b). The defender then stands upright and blocks the face of its attacker (c, d). The defender then counterattacks, reaching for its partner's mouth (e–g). The two animals then cease their interaction (h). (From Pellis, S. M. & Pellis, V. C. (1989). Targets of attack and defense in the play fighting by the Djungarian hamster *Phodopus campbelli:* Links to fighting and sex. *Aggressive Behavior,* **15,** 217–234. © 1989, Wiley; adapted with permission of Wiley-Liss, Inc. a subsidiary of John Wiley & Sons, Inc.)

is that during play fighting, they compete for access to the species-typical sexual targets, not the species-typical aggressive targets. Thus, the rat is not alone in centering its play fighting on sexual, rather than aggressive, targets.

Even though it is comforting to know that rats share features of their play fighting with other species of rodents, it could be that murid rodents are an aberration and so may not be a very useful group from which to draw generalizations about play that can be applied to other mammals. Fortunately, this is not the case. First, we can compare the play fighting seen in murid rodents across a wider swathe of rodents, and second, we can then make that comparison across a wider swathe of mammals, so as to determine the idiosyncratic features that are specific to the murid rodents and the features of their play fighting that they have in common with other animals.

The rodents can be thought of as comprising three major subgroups. As well as the mouse-like rodents (murids), there are the squirrel-like and the guinea pig-like rodents.[6] Fortunately, most squirrel-like and guinea pig-like rodents are active during the day and live in open terrain, so there have been many observations of their play in the wild. Unfortunately, there is little in the way of systematic studies of the play in these groups under standardized conditions, but there are enough data to make reasonable comparisons as to the content of their play fighting. Generally, play fighting appears mostly to involve some combination of aggressive and sexual play.[7] This does not mean, as many theorists have thought,[8] that play fighting is a mixture of behavior patterns that are derived from sex and aggression. Rather, any given play fight can involve either sexual or aggressive targeting and tactics of attack and defense. Indeed, in detailed studies of species that are capable of employing more than one type of play fighting, sequences involving one type are completed before the animals switch to the other.[9] That is, if more than one type of play fighting is available to them, then sequences of play fighting are faithful to the target initially attacked (sexual if the target is sexual, and aggressive if the target is aggressive). Thus, when comparing across rodents, it is possible to score their play fighting as sexual, aggressive, or some combination of both.[10] So, how peculiar are murid rodents? Before answering this question on the status of play fighting in murid rodents, we need to consider how to compare character states across species.

On Making Species Comparisons

Modern comparative biology is built on the foundations of a method for taxonomic classification that rose to prominence in the latter part of the twentieth century.[11] Briefly, the method involves establishing patterns of relationships among organisms so that all that are shown are the clusters of the closest relatives (i.e. clades). The questions of how long ago did they diverge from each other or how different do they look from each other are irrelevant, as all that matters is that, within a given group of creatures, some are more closely related to each other. Earlier methods emphasized overall similarity as the key criterion for judging which species should be classified together (i.e. grades). For example, consider a cow, a salmon, and a lungfish. Clearly, lungfish and salmon share many more characteristics in common than either of them do with cows – they both have a streamlined body, fins, and gills, all of which are absent in cows.

The instigator of the modern approach, Willi Hennig, realized that not all features of organisms are equally valuable in identifying patterns of relatedness. All three of our animals evolved from a fish-like ancestor and so the many fishy traits of lungfish and salmon are remnants of that earlier ancestor – that is, these are shared, primitive features that provide little information about relatedness. More useful are shared, derived features that indicate a more recent, common ancestor. As it happens, cows and lungfish share such traits. One is a peculiar construction of the nares (i.e. the external openings to the nasal passage). Salmon, as is true for archaic, fossil fish, have nares that end in a blind sac. In contrast, lungfish and cows have nares that open in the roof of the mouth. Open nares are a shared, derived trait that evolved in the common ancestor of lungfish and cows, an ancestor that diverged later than their common ancestor with salmon. The open nares are thus a trait that is informative about relatedness between these three species in a way that fins are not. The application of this approach led to the realization that even though, in overall appearance, chimpanzees resemble gorillas more than humans, chimpanzees are, in fact, more closely related to humans than they are to gorillas. Because this classification method uses such key traits to identify which members of the compared species are more closely related and so form clades, it is called cladistics. Consequently, the trees with branches linking species that are generated by this approach are called cladograms.[12]

Several computer programs are available to aid in the construction of clado-grams and some of these provide a means by which the evolutionary changes in the character states of traits across the tree can be traced.[13] Based on this methodology, the distribution of types of play fighting can be compared across a number of species of rodents from all three main subgroups (figure 3.5). What this figure shows is that the species within the murid lineage (the first main cluster at the top of the tree) overwhelmingly engage in play fighting that is based on sexual behavior. The figure also reveals, however, that some members of the other two main groups engage in mostly sexual play and that, even though the terminal branches of the tree may exaggerate either sexual or aggressive play, the ancestral state is for the presence of both forms of play fighting. Therefore, two conclusions can be drawn from this analysis – first, that the murid rodents are not aberrant in the form of their play fighting, as similar emphases on sexual play have arisen multiple times among rodents, and second, that the presence of both forms of play fighting coexisted at the foundations of the group as a whole.

The same conclusions seem to hold more widely across mammals. Unfortunately, systematic data are presently absent, but what is apparent is that both sexual and aggressive play are prevalent across a range of mam-malian groups, and in carnivorous species, this may also include play fighting that is based on competition over the body targets otherwise used in preda-tion. The presence of sexual play has probably been grossly underestimated, as it has only been recorded in cases where the animals mount one another. There are conceivably many more species like the murid rodents in which the targets attacked are precopulatory ones. Indeed, primates are probably good candidates for this sex-based play fighting.[14] However, even among some murid rodents, non-sexual play fighting occurs. For example, although most bouts of play fighting in Djungarian hamsters involves attack and defense of sexual targets, about 30% involve competing for bites at the rump, and so is aggressive play. Similarly, although most bouts of play fighting in grasshop-per mice involve attack and defense of the sides of the face, neck, and shoul-ders, which, if contacted, are nuzzled and groomed, and so represent sexual play, about 10% involve competing for gentle bites at the nape – a predatory target – and so represent predatory play.[15] A comparison of play fighting that involves competition for access to a body target, irrespective of whether that target is derived from aggression, sex, or predation, and whether the compar-

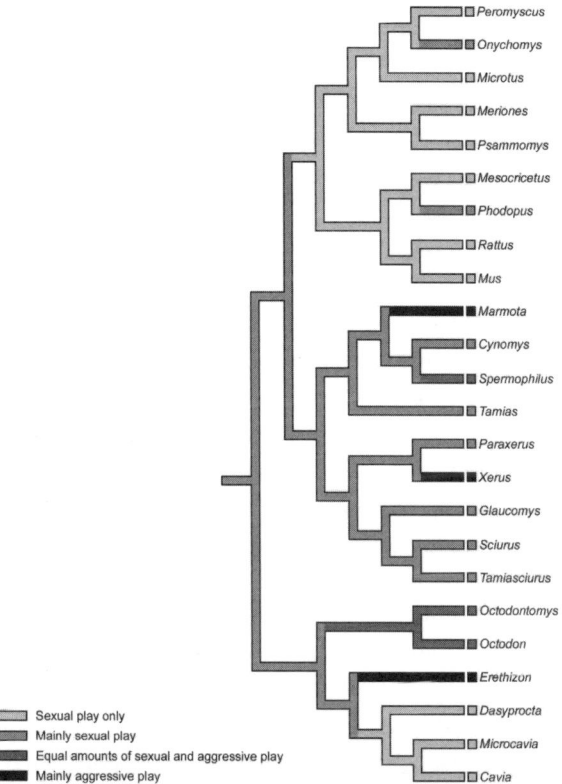

Figure 3.5 A cladogram is shown for rodents, spanning the three major divisions of the order. The clade of murid rodents includes several that we have encountered in the text, such as the rat (*Rattus*), the domestic mouse (*Mus*), the deer mouse (*Peromyscus*), the grasshopper mouse (*Onychomys*), the fat sand jird (*Psammomys*), the Syrian golden hamster (*Mesocricetus*), the Djungarian hamster (*Phodopus*) and voles (montane vole – *Microtus montanus*, prairie vole – *M. ochrogaster* and European vole – *M. agrestis*). The clade of squirrel-like rodents includes the North American ground squirrels (*Spermophilus*) and the gray tree squirrel (*Sciurus*). The other clade is composed of the guinea pig-like rodents, including the guinea pig (*Cavia*) and the degu (*Octodon*). Whether the play fighting is mainly aggressive, mainly sexual, or some combination of both, is mapped onto the cladogram. (From Pellis, S. M. & Iwaniuk, A. N. (2004). Evolving a playful brain: A levels of control approach. *International Journal of Comparative Psychology*, **17**, 90–116. © 2004, International Society for Comparative Psychology, reprinted with permission.)

ison is between or within species, indicates a fundamental commonality. For play fighting to remain playful, the competition must be curtailed by some degree of cooperation. That is, there has to be some degree of fairness in play fights.

Reciprocity and Fairness in Play Fighting

It is likely that everyone remembers a schoolyard encounter in childhood in which one individual attempted to take advantage of the situation when playing. Such an encounter may have either come to serious blows or that individual may have ceased to be an attractive play partner. With appropriate species-specific changes in the exact content of the play fighting, the same scenario appears across all animals that engage in play fighting. For this reason, play fighting is the epitome of reciprocity, since for a play fight to be playful, both partners must receive something from the exchange.[16] As we saw in chapter 2, for rats, nuzzling a partner's nape is rewarding, but it is even more rewarding if the partner offers some resistance. A partner that remains passive and allows its nape to be quickly nuzzled is a boring playmate. Conversely, an animal that offers such strong resistance that its partner never has a chance to nuzzle its nape, is also an unattractive playmate. Indeed, among juvenile rats, on the odd occasion that a play fight escalates into a serious fight, it usually arises when one animal, A, is so effective in blocking the actions of another, B, that B retaliates with a serious bite to A's rump.[17] So, for play fighting to be enjoyable, there must be some minimum level of competitiveness, but because the competitiveness must be restrained, there must also be some minimum level of cooperation. Irrespective of whether it is based on aggressive, sexual, or predatory targets, play fighting must blend competition and cooperation. Thus, play fighting in murid rodents, even though it is based on sexual targets, is not aberrant when compared to the play fighting typical of other mammals. But how do animals – rats or other species – ensure that the balance of competition and cooperation during play fighting is such as to maintain fairness in these interactions?

One way of conceiving of fairness in play fighting is in terms of the two participants having an equal opportunity to win – this is referred to as the 50:50 rule. Mathematical simulations have confirmed that for hypothetical animals engaged in play fighting, the further the win–loss ratio deviates from 50:50, the more likely it is that the animals will cease playing.[18] One problem that arises from these considerations is that some researchers report that, when play fighting, animals do not seem to exhibit any restraint in subduing their partners – rather, they play to win.[19] To resolve this problem, we need to consider two features of the 50:50 rule. First, the 50:50 rule should be interpreted metaphorically and not prescriptively. What the 50:50 rule highlights is that one partner does not completely take advantage of the situation, so that there is a reasonable chance that the fortunes of the partners can be reversed, and not that each partner has to win, literally, 50% of the play fights. Second, different species may use different avenues by which to achieve the fairness implied by the 50:50 rule. There are many different points during the playful encounter at which an individual may restrain itself from taking advantage of its partner.

Recall our discussion in chapter 2 of the problem an attacking rat faces when attempting to deliver a serious bite to the rump of an opponent. The problem is how to deliver such a bite without subjecting itself to a retaliatory bite, by the defender, on the side of its face. Figure 2.5 illustrates one solution to this problem. In this situation, the attacker uses its own flank and rump to block retaliatory strikes. However, by pushing its opponent with its flank, the attacker not only prevents retaliatory bites, but it also unbalances the defender and leaves its rump exposed to a potential bite.[20] This lateral maneuver is part and parcel of a broader principle used by attacking animals – to incorporate a defensive component when attacking so as to mitigate the threat of retaliation. That is, the attacking animal uses the tactics of attack and defense simultaneously. Similarly, when defending, defensive and retaliatory tactics have to be executed at maximum speed so as to limit the ability of one's opponent to land a bite or a blow successfully. When two animals are well matched in their fighting skills, the attacks and counterattacks may be parried so effectively that the animals do not hurt one another. Indeed, the whole encounter may be mistakenly viewed as "ritualized" – a situation in which animals try not to inflict damage on each other. That such a view is mistaken is quickly evident when one of the combatants falls off-balance, and its opponent strikes hard.[21] Play fighting need not follow such a tight coupling of attack and defense.

For many species, not only rats and other rodents, but also monkeys and birds, play partners do not incorporate defensive tactics into their attacks, and so do not limit the retaliatory capability of their partner. Similarly, during play fights, defensive actions are often performed more slowly than when they are used in serious fighting, affording one's partner the chance to succeed in its attacks and counterattacks.[22] Thus, when play fighting, these species handicap their offensive and defensive behavior, and by so doing, give their partners a chance to reverse their roles. Fairness is therefore built into the very movements performed during these encounters. Other species use different tactics to achieve fairness when playing. For example, in degus – a South American rodent that is a relative of the better-known chinchilla, which is now a relatively common household pet – the juveniles engage in play fighting that can be either sexual or aggressive. During their sexual play, these animals compete for access to the shoulders and to the side of the face of their partners. These body targets are nuzzled and groomed if contacted. When competing for access to these targets, juvenile degus appear to follow the same rules of playful combat as rats – the tactics of attack and defense are not activated simultaneously, but rather, they occur in a way that allows the partner a chance to reverse its role. However, when engaged in aggressive play, the rules differ.

As in serious fighting, aggressive play in degus involves an animal occasionally delivering a bite to its opponent's flanks. Although access to the flanks can be gained by direct assault, the opponent will typically turn to face the attacker. This results in the animals rearing onto their hind legs, grabbing each other's forelegs and wrestling, in an attempt to push or pull their partners off balance. Sometimes, from this upright position, one partner will turn and kick their opponent in the stomach, kangaroo-like, using both hind legs. The effect of such a kick can be devastating, as the recipient is hurled backwards and onto its back or side. When these types of kicks are delivered during serious fights, the attacker attempts to take advantage of the other's misfortune and delivers a bite to its opponent. This is not true for play fighting, in which the attacker will not usually pursue its off-balance partner. Indeed, the winner of the kicking contest may even allow its partner to regain its footing, approach and nuzzle the play targets without any resistance.[23] In order for an attacker to deliver a hind leg kick successfully, no matter whether it occurs during a serious or a playful fight, the attacker has to use the tactics of attack and defense simultaneously.

The degus does not, therefore, follow the rules of play typically followed by rats and monkeys. However, once the kick is delivered successfully during a play fight, the attacker will not attempt to take advantage of the situation as it would in a serious fight. Thus, there are multiple ways of maintaining fairness during play fighting. The spirit of the 50:50 rule is that partners do not take advantage of any momentary success. The partner must have a chance to gain or regain the upper hand.[24] Again, these analyses indicate that rats and their murid rodent kin are not aberrant, but follow rules that apply either in specific detail, or in their general principle, to most species that engage in play fighting.

How Much Like Sex is Murid Play Fighting?

Even for murid rodent species that engage in the full panoply of playful attack, when defense and counterattack are compared there are some striking differences. Although some of these differences can be explained by species differences in the adult behavior being mimicked during play fighting, others cannot. For example, both prairie voles and montane voles mainly use facing defense during play fighting, but whereas prairie voles more often use supine defense (which results from a complete rotation around the long axis of the body), montane voles more often use upright defense (which results from a rotation around the vertical axis of the body while standing on all four feet). Comparison of the play fighting of these two species could lead one to conclude that play fighting in the prairie vole is more complex because its predominant use of the supine defense tactic facilitates prolonged wrestling and bodily contact. Although such a conclusion is true superficially, it misses a key point. A comparison of juvenile play fighting with adult sexual encounters shows that the pattern of play varies with the species-typical differences in adult sexual behavior. During sexual encounters, adult female prairie voles are more likely to defend their napes by rolling onto their backs (turning to supine), whereas adult female montane voles are more likely to defend their napes by standing upright. That is, the most common defensive tactics in sexual encounters are also the most common defensive tactics in playful encounters. Consequently, play fighting mimics sexual behavior in a detailed manner.[25]

Rats and Syrian golden hamsters differ strikingly. During sexual encounters between adults, both female rats and female hamsters are, respectively, most likely to evade nape and cheek contact. Yet, for both species, play fighting most often involves facing defense, especially turning to supine. The most common defensive tactics in sexual encounters are therefore the least common defensive tactics in playful encounters. The mapping of play fighting onto sex in hamsters, as compared to voles, is less rigid.[26] Based on these species differences, we can add another layer of species comparison in table 3.1 – namely, whether play fighting involves a modification of the species-typical pattern of sexual interactions. The answer is that for the species compared, only two – rats and Syrian golden hamsters – show this modification. For some species, then, play fighting is more than just the precocial expression of sexual behavior. The underlying control mechanisms regulating sexual behavior have to be altered to produce the behavioral pattern present in play fighting. There is one more layer to unravel, and to understand this we need to turn our attention to the brain.

The Neural Control of Play Fighting

Let's begin with the simple question of whether the entire brain is needed to generate sequences of play fighting. Many textbooks of mammalian neuroanatomy divide the brain into two major components – the cerebral cortices and then the remainder, which is referred to as the brainstem.[27] The reason for this is that when the skull of a mammal is opened, the largest and most evident feature is the cortex. In humans, apes, and dolphins, the convoluted cortex is so large that the remainder of the brain is almost completely hidden from view; the exception being the cauliflower-shaped cerebellum that extends dorsally from the brainstem. In rats, although the cerebral cortices are not as large or as convoluted as in humans, they still fill the majority of the cranial space beneath the skull (figure 3.6). A simple way to think about this anatomical division of function is that the brainstem takes care of most of the necessities of life, from automatic processes like respiration, chewing, and swallowing, to the more variable processes that lead to finding and extracting food or other resources from the environment, while the more sophisticated processes of perception, cognition, and making connections with past memories involve the cortex.[28] Given that play fighting can be a very complex behavior, we might expect, at least for some species, that its generation requires some cortical input. This is not the case. As we mentioned earlier, rats and hamsters that have had their cortex removed surgically (are decorticated)

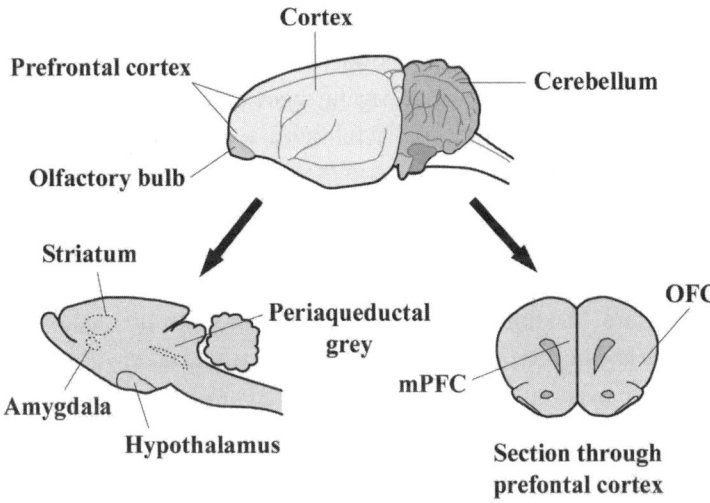

Figure 3.6 The rat brain is shown from several perspectives. In the top panel, a lateral oblique view of the whole brain is shown, with the front end facing left. Several structures are evident: the largest, and most striking, is the cerebral cortex, comprised of two lobes, one on the left and one on the right side of the long axis of the brain. At the rear of the brain is another large structure, the cerebellum, and coursing further to the rear is the spinal cord. Note that at the posterior of the brain, where the two cerebral cortices meet at the centre, there is a small triangular space and two mounds can be seen – these are the superior colliculi, brainstem structures that process visual information. In mammals with a larger cortex, such as humans, the cortices expand to cover these subcortical structures. At the front end of the brain, sitting above the olfactory lobes, is the prefrontal cortex, a structure that looms large in our story. Another area important to our tale, the motor cortex, is in the area just behind and to the right of the first vertical groove (Y-shaped line). In the panel below, and to the left, a sagittal section (length-wise along the middle of the brain) of a rat brain is shown with most of the various areas that are discussed in the text labelled, thus giving the reader an idea of the general layout of the mammalian brain. In the panel below, and to the right, is a coronal section (transverse across the brain) at the level of the prefrontal cortex, showing two areas of this part of the cortex that are discussed at length in the text: the orbital frontal cortex (OFC) and the medial prefrontal cortex (mPFC).

at birth play in a seemingly normal fashion as juveniles and young adults.[29] Armed with our schema for comparing complexity (table 3.1), and the knowledge that we have gained so far, we are now in a position to ascertain exactly what "normal" means.

Decorticated rats initiate playful attack just as often as their intact counterparts, and their playful attack follows the same, age-related modulation as in intact rats. In addition, when defending themselves, they are able to access all the species-typical patterns of playful defense, and, again like intact rats, they mostly use the tactics of facing defense. Play fights that involve decorticated rats are just as intricate, sustained, and repeated as those of intact rats, with escalation to serious fighting being rare. Together, these properties suggest that even without a cortex, such rats are capable of the reciprocity necessary to sustain playfulness (table 3.1). This suggests that many of the quantitative and qualitative features that distinguish rats from mice and voles do not require any cortical regulation. So, the two species of murid rodents on our list that show the most complex play fighting do so by species-typical changes in the regulatory mechanisms that control play in the brainstem.

These changes are likely to include modification to the upper brainstem mechanisms that are involved in regulating motivation and reward. Such a modification would lead to elevated frequencies of initiating playful attacks, changes in the thresholds set for using different defensive tactics in the motor regulatory systems, changes in using facing rather than evasive defense, and, finally, to changes in the emotional regulatory systems that enable animals to sustain more frequent and prolonged interactions while still maintaining a playful mood.[30] The motivation to engage in any particular type of activity, such as sex, eating, or play, involves neural circuits centered on an area called the hypothalamus. This is an area of the brainstem that is actually composed of several substructures, and which, as the name implies, resides beneath the thalamus (see figure 3.6). Changes in the hypothalamus, or to circuits that feed into this area, are likely responsible for two features of play fighting that indicate an animal's interest in this activity. One obvious feature concerns how frequently an animal will initiate a playful attack, but a second, less obvious one, is how likely an animal will respond defensively to a playful attack. Experimental and developmental data on rats, as well as comparisons across species, indicate that playful attack and playful defense can vary independently,[31] but given that both influence the overall incidence of play fighting, both are likely to involve neural circuits in and around the hypothalamus. The involvement of hypothalamic circuits in the motivation to engage in play fighting can also be demonstrated by using a test situation that measures the likelihood of rats to enter an arena which they have played in before versus one which they have not. The more likely the rats are to enter the play arena, the more it can be said that they are motivated to play. Manipulation of the circuits in and around the hypothalamus, using drugs that interfere with the neurotransmitters that are used in these circuits (especially a structure known as the nucleus accumbens), can alter the likelihood of entering the play arena, and this confirms the involvement of this circuitry in the motivation to play.[32]

Many of the species-typical behavior patterns that are used in a variety of situations, such as sex, aggression, and predation, are organized in the neural circuitry of the lower brainstem, especially in a structure called the periaqueductal gray area. This latter structure has been especially well mapped for defensive actions that can be used in a variety of contexts.[33] The selection of which behavior pattern to use in any particular situation usually involves regulation by structures in the upper brainstem, in particular, the basal ganglia, and especially the striatum. Disruption of the striatum leads to rats that are still able to engage in play fighting, but have difficulties in organizing their sequences of offensive and defensive actions. For example, they may begin their playful defense using one behavior pattern, but then, before completing that action, switch to another. That is, if the striatum is disrupted, it will lead to alterations and confusion in selecting the appropriate defensive tactic to be used for a particular attack.[34] Therefore, the species differences in the use of tactics of playful defense can be accounted for by differences in their threshold for activation, either by changes in the periaqueductal gray area or in the striatum. In rats and hamsters, because the defensive tactics that are most often used in play are the reverse of those that are most often used in sex, a regulatory change in this circuitry must have occurred. In comparison, in voles and other rodents, the likelihood of using tactics of defense is the same in both play and sex. Since this is a regulatory change, rather than one of actual behavioral content, the evolutionary shift from the vole-like pattern to the rat-like pattern likely involves the striatum.

Another ability that remains intact following decortication is that of maintaining fairness during play fighting. Thus, the ability to engage in reciprocal, playful exchanges is unaffected by the absence of the cortex. The amygdala, a subcortical structure located in the medial temporal lobes, has been strongly implicated in the regulation of social behavior, and, indeed, in rats, damage to the amygdala disrupts social recognition and play fighting.[35] Comparative studies of adult sexual play and juvenile social play in primates have shown that the prevalence of these play behaviors is significantly predicted by the size of the amygdala, in that the species that play the most tend to have a larger amygdala. Finally, altered function of the amygdala has been implicated in autism, a key symptom of which is an inability to reciprocate during social encounters, which leads to impoverished social play. The amygdala may be involved in play fighting either through its role in the learning and modulation of fear, or in providing a mechanism for impulse control. Excessive fear can prevent an animal from accepting playful overtures or other, amicable, social contact. Furthermore, animals with too little impulse control quickly become unattractive play partners. Either way, monkeys with amygdala lesions engage in atypical play compared to intact animals in that they seem to misinterpret playful contacts by partners. For several reasons, then, the amygdala is the prime candidate to be an important part of the subcortical neural system that regulates the reciprocity crucial to the maintenance of play fights.[36] Whether the involvement of the amygdala is the same in all species, or whether it involves different, immediate causal processes or different developmental trajectories, remains to be determined.[37] Nonetheless, it seems likely that the subcortical regulation of reciprocity during play fighting involves the neural circuits that include the amygdala.

The paucity of our knowledge of the brainstem mechanisms involved in regulating play fighting prevents us from mapping the species differences in different facets of play fighting onto the modification of specific neural circuits. At the very least, we can hazard a guess that species differences in the motivation to play (i.e. the frequency of playful attack and the likelihood of playful defense) probably involves changes in the hypothalamus/nucleus accumbens circuits, and species differences in the types of defensive actions most likely adopted involves changes in the striatum/periaqueductal gray circuitry. Species differences in reciprocity are poorly characterized at the behavioral level, and other than the smoking gun evidence for the involvement of the amygdala, at the moment, little else can be predicted. The one direction in which we are willing to conjecture is to suggest that, given the important role of the amygdala in assessing facial and bodily cues for their emotional content, then in species that show a greater reliance on using visual cues from body postures and facial gestures during play fighting (like monkeys, for example) the role of the amygdala has probably expanded.[38] At the behavioral level, species that use visual signals during play fighting to indicate that what is about to occur, is occurring, or has occurred, is or was play, have the most complex patterns of play fighting.[39]

Combining decortication experiments with cross-species comparisons provides us with a rat's-eye view of the core neural control mechanisms that are needed to generate and regulate complex patterns of play fighting. Surprisingly, all the important regulatory mechanisms that are essential for this task are subcortical. A general principle that we can derive from this analysis is that, with the advent of greater behavioral complexity, there is a corresponding addition (or modification) of an associated neural mechanism. The story derived from rats has another surprise in store for us; it reveals that even though cortical mechanisms are not needed to produce play, the cortex can modify the content of play fighting and so further increase the complexity of the behavior possible.

The Role of the Cortex

In chapter 2, we described the two forms of age-related changes in the play fighting of rats. First, the frequency of playful attack peaks at the mid-juvenile stage (thirty to forty days post-birth), and second, whereas the likelihood of defense does not change with age, the likelihood of using different tactics does. As we have already noted, decorticated rats show the same age-related changes in the frequency of playful attack and the same likelihood of defense as do their intact partners.[40] However, the age-related changes in the use of defensive tactics are eliminated in decorticated rats. This cortical effect is best seen in male rats, in which there are two changes in defensive tactics with age. The first change occurs shortly after weaning, when the complete rotation tactic replaces the partial rotation tactic as the most frequent one, and then, with the onset of puberty, there is a switch back to using the partial rotation tactic most often (see figure 2.4). Decorticate rats fail to show these switches in defensive tactics, but instead, at all ages, use the partial rotation tactic most often.[41] But that is not all: following puberty, male rats establish dominance relationships. Dominant males, when play fighting with other rats – be they other dominant males, subordinate males, or females – are most likely to use the partial rotation tactic, but subordinate males differ, depending on the partner, in what tactics they use.[42] When a subordinate male plays with another subordinate male, or a female, it will most likely use the partial rotation tactic, but when the play fight involves a dominant male, it will most likely use the complete rotation tactic. So, when play fighting with a rat of equal or lower status, the subordinate rat behaves like an adult male, but when interacting with a rat of superior status, it behaves like a juvenile! Decorticated male rats fail to modulate their style of defense with partners of different status – irrespective of whether attacked by a dominant or a subordinate, these rats will most likely defend themselves by using the partial rotation tactic.[43]

Even though rats without a cortex can generate play at the normal frequency, use all the behavior patterns typically available, and maintain the playful reciprocity necessary for play to remain playful, their play, as we have seen, is not completely normal. The cortex, it seems, is involved not in generating play, but in modulating its expression – and in rats, we have identified two forms of such modulation: an age-related one and a partner-related one. Furthermore, we have shown that the two forms of modulation are dependent on two different areas of the cortex.

The age-related modulation in playful defense is dependent on the neural circuits of the motor cortex and the partner-related modulation in playful defense is dependent on the orbital frontal cortex. Damage to either of these two areas hampers one form of modulation, but leaves the other intact. Such a double dissociation provides compelling evidence that different neural circuits are involved in these two forms of playful modulation.[44] Each of these cortical areas independently influences the brainstem mechanisms that generate play fighting. For rats, then, we have an idea of the basics of the neural circuit involved in the production of play fighting[45] – one that spans all levels of the brain, with different neural systems introducing more regulatory control over the expression of the behavior, and with the presence of neural controls from the cortex producing even more complex patterns of playful expression. But how do other rodents compare?

Of all the species for which we have quantitative developmental data as well as data on the pattern of interaction during precopulatory encounters, only rats and Syrian golden hamsters exhibit a pattern of defense during play fighting that differs from that seen in adult sexual behavior. Of these two species, only rats show the age-related changes in the types of facing defense; hamsters mostly use the complete rotation tactic, which leads to supine defense at all ages. Thus, while hamster play differs from their adult sexual behavior, no age-related modulation is needed.[46] Not surprisingly, decorticated hamsters appear not only to play as frequently as their intact counterparts, but they also do so in the normal manner, with wrestling arising from the use of supine defense.[47] Thus, for both rats and hamsters, brainstem changes are required to modify the sex-typical pattern of defense during sexual encounters into that typical of play fighting, and these remain play-typical even after the removal of the cortex. However, only in rats does this modified brainstem circuitry come under the regulatory control of the cortex. This is certainly true for the age-related modulation of playful defense that is seen in rats – given the absence of this modulation in other species, the need to regulate it with cortical control mechanisms is unnecessary. However, the story for the partner-related modulation is a little more complex.

In many species, play fighting becomes rougher with the onset of puberty, especially among males.[48] However, the degree of roughness is curtailed by the dominance asymmetry that also develops, typically, around puberty. This is well illustrated in Syrian golden hamsters. With the onset of puberty, subordinate males become increasingly unlikely to initiate play with dominants as well as increasingly resistant to respond playfully to the playful attacks of dominants. Indeed, in one highly asymmetrical pair we observed, the dominant hamster made repeated attacks to its partner's cheeks, while the subordinate squatted in a corner and failed to respond. Eventually, the dominant hamster grasped the subordinate's cheek in its mouth and dragged it around the cage, but again, to no avail – the subordinate simply curled up and allowed itself to be dragged about. It seemed that no matter what the dominant partner did, the subordinate refused to engage him in play.[49] As is the case in rats, it is highly likely that the recognition process that enables a subordinate hamster to identify its partner as dominant involves the orbital frontal cortex. However, the sequelae of recognition are very different for rats as compared to hamsters. The subordinate rat will initiate more playful attacks on its dominant partner than it will on other subordinates, and when the dominant rat initiates a playful attack, the subordinate will most likely use the supine defense. So, even if similar cortical mechanisms are involved in the modulation of partner-related behavior in rats and hamsters, the mechanism in rats has been modified to permit new responses.

Clearly, as we consider the differences across the various murid rodents examined (table 3.1), we can see that it takes many changes and additions to the neural control mechanisms regulating play fighting to go from a rodent brain that does not produce play to one that produces play as complex as that present in rats. What processes are responsible for these changes in the organization of the brain?

Is Complex Play Fighting a By-Product of Changes to the Brain and its Development?

Consider the big, heavy-set limbs of an elephant. If a species of mouse existed that had legs that were proportionately like those of an elephant, it would be reasonable to assume that in this species, such heavy-set limbs must have evolved to solve some peculiar locomotion problem. However, if we saw spindly limbs in mouse-sized animals and heavy-set limbs in elephant-sized animals, we would not be surprised. An explanation that would satisfy our observations would be that because big animals are heavier, they require bigger, thicker legs to support them against gravity, and so the differences in their body size would be sufficient to explain the differences in their limb proportions.[50] In a similar manner, the changes in the neural control over play fighting could be a by-product of some other change in body form or behavioral function. Two patterns of correlated change have been well studied and these may help explain evolutionary changes in play complexity. First, as brains increase in size in a lineage of animals, a number of changes in their organization occur that can lead to novel capacities.[51] Given that play appears to be more prevalent in species with larger brains, changes in the capacity to generate more complex play may be a by-product of evolving larger brains. Second, many lineages of animals have evolved novel patterns of anatomy and behavior by extending the juvenile period, or by retaining juvenile features into adulthood.[52] Because play is most often associated with young animals, it could be that, for species with an exaggerated juvenile period, there is retention of juvenile brain organization and so a greater likelihood that they engage in play.

If complex patterns of play fighting can arise as a by-product of changes in brain size or in the degree of juvenility expressed, then we can forget about play and concentrate on explaining how behavior changes with increases in brain size or increased juvenility. Closer inspection, however, shows that, whereas these general processes may account for some aspects of play, they are an incomplete explanation for species differences in the complexity of play fighting.

Are Bigger Brains More Playful?

Casual observation of the animal kingdom suggests that the most playful animals also have the largest brains.[53] Relative to body size, humans have the largest brain and are also the most playful of all species. All kinds of theories for such a relationship have been postulated – such as play being needed to shape and fine-tune an enlarged brain[54] – but for our immediate concerns, the reasons for this relationship are irrelevant. Rather, the primary concern here is whether the presence of larger brains in some species predicts the presence of more complex play in them. In mammals, the prevalence and complexity of play does seem to be related to brain size. However, this relationship only holds true when the average play prevalence and average brain size of orders of mammals are compared (when primates, rodents, carnivores, bats, etc., are compared as a whole); that is, mammalian orders containing larger-brained members also tend to contain species that play more (e.g. primates > rodents > bats) (figure 3.7). When smaller taxonomic units, such as families within an order (e.g. Old World monkeys, New World monkeys, apes, etc., within the order Primates), or species within an order, are compared, the pattern breaks down.[55] While these findings suggest that larger-brained species are more likely to be playful, they also indicate that having a larger brain does not automatically ensure that a species will play more, or that having a smaller brain would disqualify a species from being playful.

What these analyses indicate is that the prevalence of play and the complexity of play are not simply by-products of an enlarged brain. More convincing are comparative studies that show that enlargement of particular parts of the brain is associated with more play. For example, play is more prevalent in species that have a larger cerebellum or cerebral cortex.[56] Further, whereas an enlarged amygdala and hypothalamus are associated with more social play, they are not associated with increased object or locomotor play.[57] These findings suggest that where there are positive relationships between brain size and play, they involve specific areas of the brain and specific types of play. Finally, it should be noted that to the present, because most studies have conflated the prevalence of play with the content of play, it is not possible to assert with confidence that when a positive relationship is found that it is *play complexity*, rather than *play frequency*, that is responsible. In those studies where some attention has been paid to play complexity as a separate

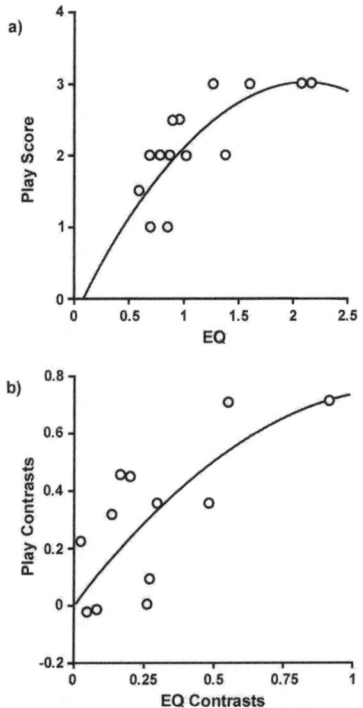

Figure 3.7 These graphs show that when compared across mammalian orders, there is a positive relationship between the prevalence of play and relative brain size. Note that this relationship holds for both raw data (A) and data corrected for the degree of relationship between orders (i.e. Independent Contrasts) (B). (From Iwaniuk, A. N., Nelson, J. E., & Pellis, S. M. (2001). Do big-brained animals play more? Comparative analyses of play and relative brain size in mammals. *Journal of Comparative Psychology*, **115**, 29–41. © 2001, APA, reprinted with permission.)

dimension of play, little convincing data has emerged to suggest that larger brains or parts thereof lead to more complex play.[58] Thus, although it may be the case that some forms of play may become more complex or more frequent with changes in the size of the brain, it seems unlikely that we can explain changes in play with changes in brain size. Given that the strongest relationships that have been found are between specific brain areas and specific forms of play, it seems more likely that the evolution of complex play has involved changes to specific brain systems and that this has been a piecemeal process.

Does Juvenility Necessarily Lead to Complex Play?

As we have already noted many times, play is most often associated with young animals, especially juveniles. But what is a juvenile? We can think of the juvenile period as the time between weaning and sexual maturity – that is, when the individual is no longer directly dependent on its parents for being fed, but is not yet equipped to start reproducing. There are many theories as to why there should be such a period of limbo in the life history of animals. One commonly held view is that young animals need a delay in the commencement of a reproductive career so as to enable them to acquire the size and skills necessary to survive and reproduce successfully.[59] For species that confront especially daunting competition, it may seem reasonable that the duration of their juvenile period is increased for them to buy the time they need to build up that competitive edge. But how do juveniles in such species use this additional time to gain in competitive ability? It has been suggested that play is the vehicle by which to achieve such skill enhancement.[60] We will defer the discussion of the adaptive value of play until the next chapter, but for our current purposes, it is sufficient to note that it is expected that species with a longer juvenile period should play more and do so in a more complex manner. Comparisons of the proportion of the developmental period prior to sexual maturity that is spent as juveniles is positively correlated with the prevalence and complexity of play fighting – and this relationship has been demonstrated for birds and two groups of mammals, rodents and primates.[61] Given the theme of this book, we illustrate this relationship for murid rodents (figure 3.8).[62]

A simple way of thinking about how to evolve a rodent brain capable of generating complex play is to view the rat brain as a juvenilized mouse brain. In this perspective, a brain that retains a juvenile character for longer is more able to generate complex play. If this were so, then, as the duration of the juvenile period is further increased, and more control features regulating play are added, this would lead to ever more complex play fighting. Certainly, the linear relationship shown in figure 3.8 makes this very plausible. Changes in play are related to changes in brain function that are, in turn, determined by changes in the relative rate of growth of the brain. One factor, then, altered rate of brain development, could account for all the variation in the complexity of play fighting. Not so fast. Is it really reasonable to explain all the specific changes in play fighting by such a general influence? For instance, why should

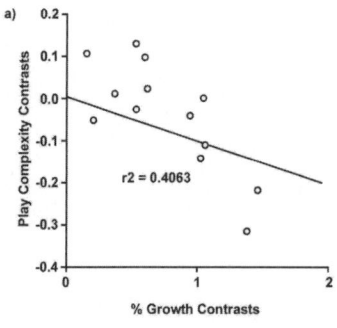

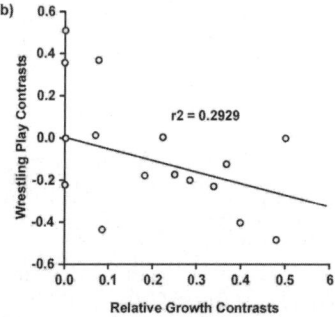

Figure 3.8 When the proportion of adult weight gained prior to birth is used, a significant negative correlation (see r-values above the graphs) is present whether a complexity score involving all facets of play fighting (see Table 3.1) is used (A), or one that simply scores whether species have a wrestling form of play fighting (B). That is, species that are more mature at birth play less. Conversely, this means that the greater the period of time spent as juveniles, the greater the complexity and amount of play. (From Pellis, S. M. & Iwaniuk, A. N. (2000a). Comparative analyses of the role of postnatal development on the expression of play fighting. *Developmental Psychobiology*, **36**, 136–147. © 2000, Wiley; reprinted with permission from John Wiley & Sons, Inc.)

the cortical control of play modulation lead to such opposite behavior in rats and Syrian golden hamsters? Similarly, why, in rats, should the motor cortex modulate age-related changes in the use of defense tactics, but not in the frequency of launching playful attacks? Maybe changes in the growth rates of the brain can explain some aspects of species differences in play fighting, but they cannot explain them all.[63]

One problem with relying on correlations is that you cannot draw conclusions about the direction of causality. That is, you cannot state with any confidence which variable is causing the other one to change, or, indeed, whether both are changing simply due to the effects of a third variable. In order to understand causality, we need to conduct experiments so that one variable can be manipulated with the assumption that, if that variable has a causal effect on the other, then that one should change in the predicted manner. If variation in play is caused by differences in juvenility, then selectively breeding rats that have a curtailed juvenile period and ones that have a prolonged juvenile period should produce, respectively, less and more playful strains of rats. Across different strains of rats there is variation in both the frequency of play fighting and in the content;[64] this suggests that there is naturally occurring variation in most facets of play that should be amenable to selective breeding.

For reasons other than studying play, a group of researchers who work on epilepsy have selectively bred rats for their susceptibility to amygdala kindling. This kindling involves delivering low doses of electrical charge, which have no apparent effect, to the rat's amygdala. However, after a number of these seemingly non-effective shocks, the rats become susceptible to seizures. Some rats require fewer bouts of amygdala kindling than others to become prone to seizure. By selectively breeding those rats that required the least kindling and the most kindling, after eleven generations, two distinct, selected lines of rats were produced – kindling-prone (Fast) and kindling-resistant (Slow). Behavioral and neural characterization of these two selected lines of rats indicated that the Fast rats were more juvenile-like as adults, whereas the Slow rats were more typically adult. For example, across a wide range of behavioral contexts, the Fast rats behaved impulsively – not unlike juveniles – and their pattern of neurotransmitter use in the amygdala retained the juvenile-typical pattern.[65] Thus, the Fast rats are more juvenilized as adults than are the Slow rats.[66] If differences in play are dependent on differences in the degree of juvenility, then the Fast rats should play more and do so in a more complex manner than the Slow rats. Indeed, Fast rats do play more than Slow rats, both in initiating more playful attacks and in having a higher probability of defending themselves. However, in terms of the content of their play fighting, both rat lines have the basic age-related changes that we have described, but both have patterns of defense that seem more adult-like – Slow evade more and Fast use the partial rotation tactic more often. Overall, then, one selected line of rats was not exaggeratedly juvenile in comparison to the other.[67]

Two conclusions can be drawn from the play of the Fast and Slow rats. First, frequency of play – how often animals seek playful encounters – does seem to be higher in more juvenile-like animals. This conclusion suggests that variation across species in *how much* they play can be explained by the degree of juvenility exhibited. Second, the content of play – the styles of defense used, and so whether the play encounter facilitates prolonged bodily contact – does not necessarily reflect the relative juvenility of the players. This conclusion suggests that the variation across species in *how* they play cannot be explained by differences in the degree of juvenility exhibited.

Explaining Species Differences in the Complexity of Play

Complexity in play fighting is built, piecemeal, by changing independent components, and, as we have seen, the evidence does not support the view that such a series of changes arose as a by-product of increases in either brain size or juvenility. Comparing the complexity in play fighting across murid rodents and mapping that complexity on a cladogram further supports the piecemeal nature of the evolutionary changes in play fighting (see figure 3.9). Figure 3.9 shows that the ancestral murid rodent, seen at the base of the tree, had a pattern of moderately complex play fighting, and that in some lineages the complexity of play fighting has increased, while in others it has decreased.

Furthermore, other aspects of the biology of some of those species reveal that many of those with the least complex play have the largest brains and the longest period of post-natal development. From these comparative analyses, two conclusions can be drawn. First, the neural changes needed to alter the complexity of play are capable of evolving in a piecemeal fashion. Second, the increases in the complexity of play fighting cannot be simply explained as by-products of increases in brain size or juvenility.

The comparative analyses also tell us something important about the adaptive value of play fighting. That some lineages have reduced the complexity of their play suggests that this complexity is only produced or retained when there is some value in doing so. The value of evolving and maintaining complex patterns of play fighting must involve benefits that convey improved survival and reproductive success. But what are these adaptive benefits, and how can such benefits help us explain the specific content of the play of any

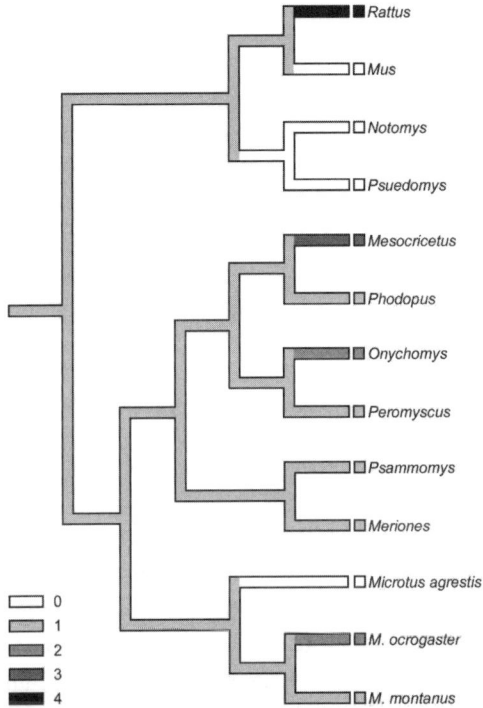

Figure 3.9 The degree of complexity in play fighting is mapped onto a cladogram of the murid rodents. The insert represents play complexity with the score of 0 indicating little or no play and a score of 4 indicating very complex play. Note that although the base of the tree indicates that the ancestral murid rodent had moderately complex play, the terminal branches indicate that some lineages have increased the complexity of their play, whereas others have reduced the complexity of their play. These findings suggest that although, in some lineages, it has been advantageous to become more playful, in others, it has become advantageous to reduce play. (From Whishaw, I. Q., Metz, G., Kolb, B., & Pellis, S. M. (2001). Accelerated nervous system development contributes to behavioral efficiency in the laboratory mouse: A behavioral review and theoretical proposal. *Developmental Psychobiology*, **39**, 151–170. © 2001, Wiley; reprinted with permission from John Wiley & Sons, Inc.)

given species? Again, rats provide a window into this issue, and once we understand the situation in rats, we can expand our horizons to other murid rodents and then to other mammals.

SO, WHAT DOES PLAYING BUY YOU?

Given that play fighting, along with other forms of play, is most often seen in young animals, and young animals are in the process of becoming competent adults, much study and effort has been devoted to trying to understand how playing as juveniles benefits them as adults. This is the most frequently asked question about play – its delayed benefits. Of course, although it is natural to think of an immature animal as being an incomplete adult, it is also the case that, at every stage of development, animals must be competent survivors, otherwise they would never become competent adults. Therefore, some theorists have noted that because play may serve to make juveniles themselves more competent, the benefit might be immediate. Furthermore, for some species, play continues well into adulthood, and because adults are already fully mature, and so are not preparing for maturity, the benefits for play in adulthood must necessarily be immediate.[1] There are, no doubt, likely to be both delayed and immediate benefits of play, but given the diversity of play, it is also likely that, depending on the type of play present and the species involved, the benefits that predominate would vary.

Let's review the evidence that suggests there are both some delayed and some immediate benefits to play fighting. We will tackle this story in bite-sized chunks, using what we know about rats as our guide. In this chapter, we will characterize the delayed benefits of play fighting, and, in the next chapter, the immediate benefits of play fighting. Understanding these delayed and immediate benefits in rats will be facilitated with a little help from other species whose play is at least as complex as that of rats. Then, in chapter 6, we can place the functions of rat play in a broader, comparative perspective, and so develop a model capable of explaining the variation in play to which we have been alerted in previous chapters.

We can begin our journey by making what would appear to be a safe prediction: because play fighting in rats involves mimicking sexual behavior, deprivation of play fighting in the juvenile period would prevent them from practicing sexual skills and so likely lead to sexual incompetence in adulthood. A large body of literature unequivocally shows that, if deprived of peer–peer play as juveniles, rats, in adulthood, will perform poorly sexually.[2] Given the mind-set of the 1960s and 1970s, and the laboratory procedures available then, these studies mainly involved a comparison of male rats that had been reared socially with ones that had been reared in isolation, that were then both exposed, separately, to a sexually competent female.[3] It was found that a play-deprived male did not orient appropriately to a receptive female, and that this led to such aberrations as mounting the female's head. Even when mounting from the rear, a male could fail to adjust his movements to those of the female and so, if she moved, would veer off her body, rather than maintain the mount. Two questions that arise from these studies are whether the same deficits are present in other species when deprived of juvenile play experience, and what is actually "learned" from the play experience. As we will discover, the answers to these two questions are interconnected.

Unfortunately, aside from the studies available on rats, there is a dearth of data on the role of juvenile play fighting on subsequent sexual performance in other murid rodents. The only other rodent that has been studied in anywhere near the necessary detail is the guinea pig, a non-murid rodent (see figure 3.5). When a male guinea pig courts a female, he has to sidle up to her and perform a series of steps and body waggles that has been described as "dancing the rhumba." During the juvenile period, young males direct these movements towards their siblings and mothers. Failure to obtain such experience leads to adult males that cannot organize these movements into the correct sequence and appropriately orient their movements toward a female.[4] Experiments with rhesus monkeys have yielded similar results. Male rhesus monkeys that have been reared in social isolation fail to mount a sexually receptive female correctly from the rear, but instead will mount her from the

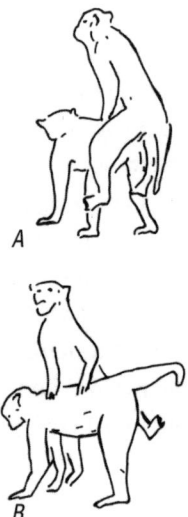

Figure 4.1 A male rhesus monkey, reared socially, is shown mounting a receptive female from the rear. When mounting, the male adopts a double clasp of the female's ankles (A). In contrast, a male reared in social isolation often fails to mount from the rear (B). Even when it does, it is unlikely to perform the ankle clasp. (From Mason, W. A. (1960). The effects of social restriction on the behavior of rhesus monkeys: I. Free social behavior. *Journal of Comparative & Physiological Psychology*, **53**, 582–589. © 1960, APA, adapted with permission of W. A. Mason.)

front or side; even if a male does mount correctly from the rear, he may not use the normal foot clasp, whereby he grips the female rhesus monkey's ankles with his feet (figure 4.1). For rhesus monkeys, we also have some good, descriptive data for females, which similarly show that a female that has been reared in social isolation will fail to present her rump appropriately to the male or will move inappropriately as the male begins to mount, which can lead to a failed attempt at copulation.[5]

Sparse though it is, the comparative data suggest that a widespread phenomenon exists – playful interactions during the juvenile period provide an important substrate for animals to gain sexual competency. However, the content of play does not always map onto specific sexual behaviors, and so the role of play in the development of sexual competency is obscure. For example, in guinea pigs, the courtship movements do not involve competitive play wrestling, and although in monkeys there are clearly interactions that can be labeled as play fighting, there is a good deal of mounting behavior that also occurs either together with, or independently of, play fighting. But as we have already seen, play fighting in rats also involves competition for bodily targets that are contacted by the male during sexual encounters. Further, in both young guinea pigs and monkeys, the recipient of the courtship movements or the mounting will take evasive action, which adds a competitive dimension to the interaction. What all these playful interactions among different species have in common is that bodily targets, postures, or orientations, need to be achieved in a context in which the partner is, to varying degrees, uncooperative. In such a context, it seems reasonable to assume that what is being learnt from the experience is how to make compensatory movements in relation to the movements of one's partner. This suggests that what is gained during juvenile play fighting is a refinement of the tactics needed for a successful sexual union. The same may be thought to apply to other skills needed by adults, such as fighting and predation.

Juvenile Play and the Rehearsal of Action Patterns

A commonly expressed view of the function of play fighting is that it serves as a means by which juveniles practice the combat skills that they may have occasion to use as adults.[6] Indeed, aficionados of nature programs on television will have seen and heard that in play fighting, animals practice fighting skills and that, in general, play is a rehearsal for many of the skills that one needs in adulthood. The argument goes something like this: during play, juveniles can rehearse the tactics of attack and defense that would otherwise be too dangerous for them to use until they were actually needed in adulthood. This contention is supported by the observation that, across a wide range of species, both play fighting and serious fighting are more common in males than in females.[7] But before we can evaluate this influential explanation for the function of play fighting properly, we first need to make a digression so as to understand both the language used and the nature of the evidence needed.

What is Meant by Function?

In the normal parlance of evolutionary biology, especially as applied to fields that use evolutionary logic to explain behavior (e.g. animal behavior, behavioral ecology, ethology, evolutionary psychology, etc.), the function or adaptive value of that behavior is the benefit conferred to an animal, in terms of its survival and reproduction. However, any particular behavior may have many benefits, so which benefit constitutes the function and which can be considered as merely incidental? One solution is to reserve the term "function" for those benefits for which we have evidence that the behavior producing those benefits has been acted upon by natural selection.[8] From this point of view, how behaviors generate their associated benefits needs to be examined in detail, providing the evidence required to identify the role that natural selection may have played in fitting the behavior to its presumed function. For example, if the play fighting of rats has the function of improving courtship skills, then play fighting should be organized in a way that makes that outcome highly likely. That is, the design of the behavior – or any other biological trait, for that matter – should match the presumed function(s).

An adaptation, then, may be viewed as a trait that, because of a history of past selection, has a specific function. Evidence of design of that trait is evidence of selection in the past and does not necessarily need to confer a currently adaptive benefit.[9] An alternative is to take a non-historical approach to function, that is, to accept that any current benefit that improves survival and reproduction is a function or an adaptive value. In some ways, such an agnostic approach makes sense, because instead of worrying about the results of past selection, we can focus on what is important for a living animal in the present and how that may impinge on its survival and reproduction.[10] Nevertheless, given that evolution is an historical process, many current functions are likely to have an imprint on them from the past.

Consider bird feathers. The most likely, original function of feathers – which were initially small and fluffy – is that they were a modification of the reptilian scales on skin and were used to regulate body temperature. Only later were feathers co-opted for use in a novel function, that of flight.[11] Imagine being a biologist back in the time of the dinosaurs and noticing that some arboreal species glide from tree to tree. Closer inspection reveals that the downy feathers covering their limbs provide some air resistance. As a result, the proto-birds that have a little more fluff stay airborne longer, can reach a neighboring tree higher up on the trunk, and so remain further out of the reach of ground-living predators, than ones with less fluff. Thus, being able to glide further and higher confers a survival advantage on fluffier proto-birds. But note that, at this incipient stage in the development of feather-based flight, there is little to no evidence that selection had an influence on fluffy feathers in the past to improve, specifically, gliding ability. Rather, all that is required is that, in our population of feather-covered dinosaurs, some are slightly better endowed (with thicker or longer feathers) than others. Those that by chance glide better survive longer, and, if the variation in feather covering has a genetic basis, they will pass this trait onto their offspring. That, by definition, is natural selection. So, even without a traceable history of selection, a current benefit can be considered a function in that it confers an advantage that actively affects reproductive success.[12]

Whether or not there is an historical trace to current functions, it should be the case that correlating variation in the trait within a population with subsequent survival and reproduction would reveal that those that are better endowed with the trait should do better. Similarly, if a trait is hypothesized to perform a particular function, then experimentally manipulating that trait should affect the bearer's survival and reproduction. For example, male widow birds attract females by displaying their long tails. By cutting and pasting tails, researchers demonstrated that males with artificially lengthened tails became more attractive to females, whereas those with artificially shortened tails became less attractive. Given the greater opportunity for natural selection to influence the form of a trait, the use of design features to determine whether the fit between structure and function supports a particular functional hypothesis is more likely to be useful if the trait in question has had a longer history of past selection.[13] One way in which to determine whether a trait has been acted upon by past selection is to see if different facets of that trait are distributed among a number of related species. The more ancient the split between existing species, the greater the likelihood that, if they share features of the trait, they probably do so because those features arose in the past – such as the vertebral column present in salmon, lungfish, and cows (see chapter 3, endnote 12 and associated text).

As we compared the play fighting of murid rodents, it became increasingly clear to us that there were many variations present – no play, rudimentary play, complex play, and much that was in-between. No doubt within that variation there are many benefits associated with playing – some are incidental benefits and some are functional, with some of those functions having

been shaped by natural selection for a prolonged period of time. Other benefits are incipient, having a current contribution to survival and reproduction,
but with little evidence of past selection. We suggest that such diversity
requires an eclectic approach to the study of function, including both
historical and non-historical methods.[14]

Applying Functions to Play

Animals that engage in more play fighting during the juvenile period practice
the tactics of attack and defense more often, and this should lead to better
combat skills as adults. As we have already noted, this explanation seems so
intuitively obvious that it is a favorite of practically all nature documentaries.
And, even though play fighting in rats has no demonstrable practice effect on
serious fighting,[15] their sexual performance is affected. Play fighting in rats
rehearses sexual behavior and the sexual behavior is under par when individuals are deprived of juvenile play experience. So, by expanding the domain of
what is included in play fighting and the domain of the behavior rehearsed,
the general principle seems to hold – play fighting experience improves the
performance of behavior patterns in adulthood. Karl Groos would be pleased
– after all, in 1898, this was the core of the explanation that he proposed for the
existence of play. He hypothesized that animals have juvenile periods precisely so that they can play and so refine the skills that they need as adults.[16]
The data on rats seem even better than one could ask for – not only are behavior patterns that are practiced in play more refined when rats use them in sex,
but they are also improved when performed in other contexts important for
survival in adulthood.

For example, when rats find a small piece of food, they hold it in both
forepaws, lean back onto their hind feet and eat the item. Rats, being social,
will approach a partner that is eating a piece of food, reach over to sniff the
item, and then attempt to rob them of it. The rat holding the food will evade
this by pivoting around a vertical axis along the length of its body ("dodging")
and so laterally move its mouth away from the other rat (figure 4.2). Males
and females differ in the anatomical location of the point of pivoting. Females
pivot around a point near the base of their pelvis, whereas males pivot around
a point closer to their mid-body (figure 4.3). These sex differences primarily
arise from how the nervous system organizes movements, and are not

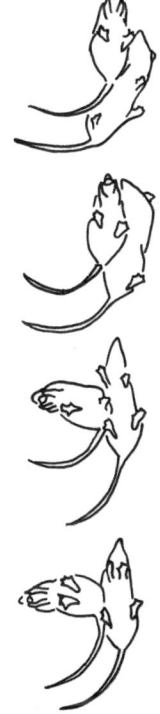

Figure 4.2 Food robbing and dodging are shown in rats. The robber, on the right, approaches the mouth of the rat that is eating. This leads to the rat that had been eating swerving laterally away from the robber. (From Whishaw, I. Q. (1988). Food wrenching and dodging: Use of action patterns for the analysis of sensorimotor and social behavior in the rat. *Journal of Neuroscience Methods,* **24,** 169–178. © 1988, adapted with permission from Elsevier.)

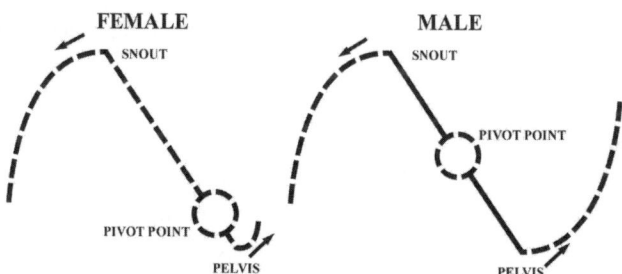

Figure 4.3 Schematic drawings show the trajectories of the snout and the pelvis during dodging by a female and male rat. Note that the pivot point in the female is closer to the end of its pelvis, whereas in the male, it is more centrally located. Because of this, males make a larger excursion of the pelvis. (From Pellis, S. M., Field, E. F., Smith, L. K., & Pellis, V. C. (1997). Multiple differences in the play fighting of male and female rats. Implications for the causes and functions of play. *Neuroscience & Biobehavioral Reviews,* **21,** 105–120. © 1997, reprinted with permission from Elsevier.)

by-products of any differences in body size or shape. In fact, these differences are not limited to dodging, but extend to all actions that require coordinating movements of the forequarters with movements of the hindquarters. Also, when rats dodge away from a male robber, a male defender will end its dodge so that its rump opposes the robber's face, whereas when a female dodges against a female robber, it will end its dodge with its rump facing the robber's mid-flank area. Dodging in rats not only occurs when a rat attempts to protect the food in its mouth, but it also occurs when it evades nape contact during sexual encounters and when play fighting.[17] Armed with these facts about dodging in rats, we can now ask what happens to a rat's ability to dodge when it is reared in social isolation, during the juvenile period.

When tested in the food robbing and dodging paradigm, males and females reared in social isolation are not only able to perform dodges, but they also do so in their sex-typical manner. No practice is needed! However, a male reared in isolation will fail to orient its rump to the robber's head. This is a difficult task for the male, because it has to modify its movements to compensate for those of its opponent so as to maintain the appropriate orientation; failure to make these compensatory movements would lead to the dodge ending in a female-typical orientation to the mid-flank. Naturally, because the head is a smaller target than the body, the male has to make more subtle adjustments to its movements to coordinate with the movements of the robber.[18] This finding is consistent with the view that play is not necessary for animals to develop particular, species-specific action patterns, but rather, it is important to refine how to use those behaviors to their best advantage.[19]

If the function of play fighting is to refine motor patterns so that their execution is more effective in the adult contexts in which they are later to be performed, then, from an historical perspective, you would predict that the behaviors performed most often during play fighting would be those that would benefit the most from practice. But this does not appear to be the case for rats. During play fighting, the most commonly performed defensive tactics are those of facing defense, especially those which lead to a supine position; yet in sexual encounters, the most commonly performed defensive tactic is evasive defense.[20] So, the most frequently rehearsed behaviors are not those most often used in adulthood. Nonetheless, from a non-historical perspective, we can conclude that practicing evasion contributes to an enhanced performance of these tactics in adulthood. However, this focus on current function, irrespective of past selection, leaves us dissatisfied, because the proposed function – the practice of adult behavior patterns – fails to provide an explanation for the occurrence of the majority of the behaviors performed during play. This is true for both the rehearsal of sex and the rehearsal of combat. For example, during juvenile play, male sheep use a variety of behavior patterns that are typically used during sex, and, consistent with the practice function, if the sheep are isolated shortly after weaning (fifteen weeks after birth), their adult sexual performance is impaired. However, young sheep play, most often, earlier in development (five to six weeks), and, if they are isolated during this time, there is no effect on their adult sexual performance.[21] Thus, the practice function of play may explain some aspects of the play experience, but it does not explain why the frequency of play varies with age in this way.

One could argue that what needs to be practiced most is not that which is used most often in adulthood, but rather, that which is the most difficult to execute. If this were the case, then we should expect that, in species where play fighting supposedly practices serious aggression, it should be more difficult to execute behavior patterns that are practiced the most. This should be especially true for species that use lethal weapons during fighting; in these cases, failure to execute species-typical maneuvers effectively could lead to dire consequences. However, the few data available do not support this expectation. For example, many ungulates that have horns (e.g. goats, cows) or antlers (e.g. deer, elk) engage in sparring during play fighting, which involves locking horns or antlers and pushing, but they rarely, if ever, engage in the tactics that are the most difficult and dangerous to execute in serious fighting, such as banging their heads together and so clashing their horns or antlers.[22] Therefore, during play fighting, the easiest tactics are the ones most practiced. Some ungulates that do not have such formidable weapons systems, such as pigs, do seem to use the full suite of fighting tactics during play fighting.[23] What is more, from our own observations of play fighting in juvenile Visayan warty pigs, we can attest to the fact that they do not show restraint during the execution of fighting tactics. These juveniles integrate offensive and defensive movements into their play fighting so as to gain the advantage over their partners, as they do in serious fighting. But, as we discussed in chapter 3, for play to remain playful there must be restraint. In the case of these pigs, the restraint appears to come after one partner has successfully overbalanced the other. If the downed warty pig squats into a submissive position, the "victor" will not press the attack.[24] Thus, it is possible that, for some species, the tactics of fighting may receive the appropriate amount of practice to influence subsequent combat performance.

One of the most detailed studies to date of the practice hypothesis has been conducted by Linda Sharpe, with meerkats, a species of small, group-living, African mongoose. In this species, both males and females are aggressive as adults. For meerkats, aggression is instrumental in gaining dominance within the group, and improved dominance status has a demonstrable influence on reproductive success. The play fighting within and between litters in free-living bands of meerkats was studied in the same individuals from early infancy to adulthood, and then, in adulthood, their success in serious combat was assessed and their position in the band's dominance hierarchy determined. The amount of play fighting in which these individuals engaged did not affect their subsequent wins in either play fights or serious fights, nor did it significantly affect any individual's eventual status. That is, meerkats do not practice fighting skills in a way that influences later fighting success or gains in dominance.[25]

Similarly, studies contrasting play fighting or predatory play with subsequent predatory skills have, generally, also been unsupportive of the practice hypothesis, whether the studies have involved correlations, as in Sharpe's study, or experimental manipulation.[26] Of course, most of the species studied so far have not been evaluated as to whether they engage in a form of play fighting that emphasizes cooperation in the manner in which the tactics of attack and defense are integrated, like rats, or in a manner that emphasizes fighting-like integration, like pigs. For these reasons, it is uncertain whether the organization of play fighting is designed so as to be able to practice combat skills. Perhaps, if the right species is used for the right behavior, then experimental evidence for the practice hypothesis may be found.

Despite the lack of direct evidence, some data do exist, especially for rats and monkeys, which show that a lack of juvenile play experience leads to reduced competence as adults in the execution of several behavior patterns. Even though lateral evasion in rats is their least used tactic during play fighting, this maneuver would be performed many hundreds of times during the course of the juvenile period, and may explain the better performance in dodging by rats with playful experience. Two conclusions seem clear from these studies on rats – the absence of the opportunity to engage in play fighting during the juvenile period leads to less well coordinated defensive food dodges and to incompetent sexual mounting in adulthood. Similarly, rhesus monkeys reared without the opportunity for playful interactions with peers are sexually incompetent, in that they fail to mount females correctly if male, or to orient to males correctly, if female. Thus, the practice hypothesis is partially successful in explaining at least some features of play fighting. Given the poor showing of most of the adaptive hypotheses of play,[27] one that is partially successful is not so bad. But before we toss our hats in the air and shout "hurrah," let's look at the evidence a little more closely.

The poor sexual performance of monkeys that have been reared in social isolation could arise because their physical skills (i.e. motor performance), social skills (i.e. the ability to interpret social cues correctly), or cognitive skills (i.e. the ability to assess the nature of the demand of the task at hand) are impaired. The practice hypothesis, then, needs to be broadened, so that the practice arising from peer–peer playful interactions provides what is needed to refine any or all of these skill sets. A clue as to what this broadening involves is provided by experiments with monkeys. When rhesus monkeys that have been reared in isolation are presented with a carved, wooden model of a monkey standing on all fours, the males are quite capable of mounting the model from the rear and using the foot clasp correctly (figure 4.4). Similarly, female monkeys that have been reared in isolation can position themselves in front of the wooden model, present their rumps, and remain stationary in the appropriate manner.[28] It is not that these isolates are impaired in their motor, social, or cognitive skills – rather, they are frightened by a live partner, and so they overreact to its slightest movement, and thus fail to coordinate their movements effectively with those of their partner. In contrast, because the wooden model is immobile, and does not emit visual or vocal signals, it is not threatening to them; they are not as fearful of the model and so are able to perform the behavioral sequence properly. This suggests that the play experience may refine the ability to deal with potentially threatening and stressful situations more adequately, and that this, in turn, allows the animals to bring their motor, social, and cognitive skills to bear more effectively on the problem at hand. The evidence from the effects of play deprivation can be re-examined in this light. Our contention is that this alternative hypothesis is plausible, and that play experience functions to calibrate the stress-response mechanism so

Figure 4.4 A male rhesus monkey, reared in isolation, is shown mounting a wooden model of a female monkey. Not only is the mount from the "rear" of the model, but the monkey is also using the double ankle clasp. (Drawing from a photograph from Deutsch, J. & Larsson, K. (1974). Model oriented sexual behavior in surrogate-reared rhesus monkeys. *Brain, Behavior & Evolution*, **9**, 211–226. © 1992, Karger; adapted with permission of S. Karger AG, Basel).

as to allow for the more effective use of available skills (of all kinds). Before examining that evidence, however, we should reconsider the isolation experiments in more detail.

Is It Really the Deprivation of Play that Causes the Deficits in Animals that have been Reared in Isolation?

If the goal of rearing juveniles in isolation is to study how the lack of opportunity to engage in play fighting affects development, then we have a problem: isolation also deprives them of the opportunity to engage in a host of other forms of social behavior. It should be borne in mind that, for a social animal, being alone is a very stressful experience. It is no accident that solitary confinement in prisons is used on humans as a form of punishment. Given that an animal is deprived of so much when it is reared in isolation, we cannot really be sure that it is the absence of play fighting experience alone that is the critical factor in producing the subsequent deficits in behavior.[29] The case for the critical role of social play in the juvenile period to later development began being built by Dorothy Einon and her colleagues in the 1970s and 1980s in a series of experiments on rats and other rodents.

First, this research demonstrated that, in rats, not only is sexual performance impaired by the lack of play fighting experience in the juvenile period, but also that it leads to myriad social and cognitive deficits. Second, various experimental procedures were used by Einon to pinpoint whether social play – which in rats primarily involves play fighting – is, in fact, the critical experience that is missing when rats are reared in isolation.

To begin with, family groups of rats were observed. This re' ealed that, over the course of the day, rats engaged in play fighting for abo' .t an hour. Armed with this information, Einon housed the juvenile rats se' arately – in isolation – but then gave each of them daily contact, for one hour, with a partner. Not surprisingly, the rats played for much of that hour. This novel departure from the experimental protocol that is usually employed for isolation experiments revealed that these rats did not suffer from the usual cognitive deficits in adulthood. Of course, the hour of social company that was provided may have been more critical to this result than the play fighting itself. Therefore, a different approach was needed to determine whether it was general social contact or specifically play fighting that was the critical missing factor.

Individual juvenile rats were housed with a single, adult female. In general, adult females play very little, and, when they do play, they tend to avoid doing so with juveniles. Thus, a juvenile growing up with an adult female will have had the experience of social company, including social grooming and huddling, but it will have had little to no experience of play fighting. Einon found that, when adult, these rats showed the same cognitive deficits as those reared in complete isolation! In contrast, when reared in social isolation, house mice, hamsters, gerbils, and guinea pigs show none of the cognitive deficits that arise in rats.[30] This comparative finding is interesting because, as we showed in chapter 3, of all the rodents analyzed, only rats seem to have the specialized cortical mechanisms that modulate play fighting, and it is the cortex that is primarily involved in the various cognitive tasks tested by Einon and her colleagues. Additional studies, not only by Einon's group, but also by many laboratories around the world, have further supported the conclusion that the critical experience that is missing in rats that have been reared in isolation is playful fighting.

If juvenile rats are reared next to another juvenile, but are separated by a wire mesh, so that they can see, smell, and sit next to one another, this is insufficient to protect them from the deficits associated with being reared in isolation.[31] It has also been established that the critical time for receiving this experience is the juvenile period – not before or after.[32] Furthermore, the deficiencies of rats reared in isolation during this period are chronic. Rats that are reared separately from social partners, but are then given several weeks with them, still exhibit the same deficiencies,[33] which indicates that it is the lack of past experience that is the culprit, not the current lack of social contact. Unlike prolonged isolation, short-term social isolation, ranging from a few hours to about a day, over many days during the juvenile period, does not lead to chronic deficits. However, short-term isolation does stimulate an increase in play fighting when the isolate is reintroduced to another juvenile that has been housed under normal conditions, in a group. The play fighting between such a pair is the same as that which occurs for group-housed animals and is preferentially engaged in relative to other social behavior, such as social grooming, social investigation, or huddling.[34] The quality of the play partner's response is also important for the playful experience. When a juvenile rat is offered the choice between an active, but drugged, non-playful partner, versus one that can interact normally, it seeks out the playful partner. Indeed, for a rat, being forced to interact with non-responsive partners appears to be downright aversive.[35]

Long-term social isolation in the juvenile period not only leads to deficiencies in behavior and cognition, but also to changes in various brain mechanisms. For example, being reared in social isolation leads to a decrease in the density of the cellular receptors in striatal neurons that respond to the neurotransmitter, dopamine. This alteration is not reversed if the rat is then socially housed after the juvenile period. And it is not only the isolate that is affected; normal juveniles when housed for a prolonged period with a drugged, non-playful peer also have reduced densities of dopamine receptors, as low as those of the isolated animals.[36] Thus, having a partner, even an age-matched one that is physically present, is not enough to offset the effects of isolation – the pair need to be able to interact actively. These data provide further support for the conclusion that it is the lack of opportunity to engage in play fighting, and not a lack of general social behavior, that is the critical deficiency in animals that have been reared in social isolation during the juvenile period. Even though the absence of play fighting is an important component of the missing experience in isolated juveniles, there are doubtless other forms of social contact that may have important roles, and these need to be carefully evaluated.

When considering the roles of particular experiences during development, we need to be cautious in assuming that, because removing A leads to X, then

A has a direct causal effect on X. Development can often take circuitous routes.[37] For example, in free-living conditions, black-headed gulls nest in dense flocks in open, flat areas and the distance between the nests is only one or two gull body lengths. Shortly after hatching, the young gulls from all these nests cluster together, forming a large, dense flock. With adults milling around the circumference of this flock, there is a measure of protection from predators. A characteristic of these large gatherings of immature gulls is that they squabble: they peck and threaten one another. As the young grow, the rich repertoire of body signals and vocalizations that are used for communication as adults gradually mature. What is pertinent to our discussion is the observation that, if the young gulls are denied the opportunity to squabble, the development of these signals is retarded or even incomplete. A logical conclusion is that during this squabbling, the gulls are practicing the different components of the signals and it is the feedback from their opponents that helps shape the final form of the signal. Not so. As it happens, the squabbling between these young gulls elevates their circulating levels of testosterone,[38] and it is this extra testosterone that facilitates the development of their signals, probably by promoting the development of the relevant neural circuits. When young gulls that had not been housed socially were treated with injections of testosterone, they not only developed the signals in the absence of squabbling, but they also did so prematurely – that is, at a younger age.[39] Thus, the social squabbling between the young gulls has an indirect effect on the development of their signaling behavior.

In the same way, it is possible that the play fighting experience may indirectly facilitate the development of important skills. For example, when rats are raised in enriched environments – that is, in enclosures that contain multiple animals, and have large areas in which to run and climb and a wide assortment of objects to explore – they are generally more active. Because young rats playing will stimulate others to join in, the increased activity may engender more play, and, in turn, more activity.[40] Increasing the opportunity an animal has to engage, actively, in the world may be an indirect way by which playing with peers exerts an influence on the development of an assortment of skills. Alternatively, the experiences derived from play fighting itself may be the direct, causal agents for the development of those skills. Given the labyrinthine nature of the evolutionary and developmental patterns so far documented for play, we believe that there is evidence that both types of processes – direct and indirect – are involved. To begin with, we suggest that one effect of play fighting experience is that it conditions the emotional capacity of the rat in such a way that it is able to use all aspects of its behavioral toolkit (motor, social, and cognitive) more effectively. This means that we need not look for how play experience affects the development of specific skills. This eliminates one of the enduring sources of argument about the function of play – does it promote the refinement of motor or social or cognitive skills? The answer to this question is yes, it does – but indirectly, through a mechanism that affects all of them.

Being Better at Coping with Stress

In a scene from Alfred Hitchcock's classic 1963 film *The Birds*, the main character, Melanie (Tippi Hedren), enters a room. All is initially quiet, but then she notices a big hole in the ceiling and roof, and sees dozens of birds roosting. Her face shows panic as she turns and then attempts to open the door of the room. This noise stirs the birds and they attack her *en masse*. As they swoop down and peck her, she becomes less and less able to carry out the simple task of turning the door handle so as to escape. In this highly fearful and stressful situation, her capacity to carry out a well-practiced action is impaired.[41]

Just as in people, when rats that are stressed are placed in a situation that requires them to perform a motor task, even one in which they are well rehearsed, their ability to execute the required actions deteriorates. For example, when naïve rats are given unshelled sunflower seeds to eat, they are initially clumsy – they turn the seed, reorienting it in their paws, stripping off a piece of shell here and there. However, after some experience, a rat becomes competent at this task. It can grab the seed, orient it so that it is held lengthwise to its teeth, and then deliver a small bite to one end of the seed. The rat then turns it again, and delivers a small bite to the other end. In this manner, the rat can cleave the shell into two neat pieces, and then remove the seed and eat it. It only takes a few days for a rat to become proficient at this task.[42] In an experiment conducted by Karen Dean, while a post-doctoral researcher in our laboratory, adult male rats were trained in this task until they all could produce two nice, clean, half shells. They were then tested next to either dominant or subordinate males, which were housed on the other side of their cage, and separated from the trained rats by a wire-mesh partition. Regardless of the status of their neighboring rat, the dominant males that had been trained were able to strip the shells of the seeds proficiently. For subordinates, however, the pattern was somewhat different. If the neighbor was also a subordinate, then the subordinate would continue to produce two clean, half shells, but if the neighbor was a dominant, proficiency went out the window. The rat would orient and reorient the shell, and would then start taking strips off, leaving piles of shell fragments. It seemed that the subordinate found that being next to a dominant was a disconcerting experience, and in being stressed, its motor competency was eroded.[43] Systematic studies by Gerlinde Metz have clearly demonstrated that when rats are stressed, not only is their skilled motor performance – such as shell hulling and reaching for food – compromised, but so are simple, unlearned, automatic actions, such as walking. When stressed, rats walk with a stilted gait, with their limbs held more stiffly.[44] Other researchers who have investigated the effects of stress in rats using cognitive tasks have also found clear evidence that the cognition of such rats is impaired.[45]

Common symptoms of these stress effects on a rat's motor and cognitive performance are a reduced fluidity of movement, a tendency either to underreact or overreact, a delay in taking action, and so on. Rats that have been socially isolated during the juvenile period show all these quirks, just as the overanxious monkeys that we discussed earlier did when trying to mount sexually receptive females. When rats reared in social isolation encounter unfamiliar rats, they are more likely to be involved in fights, and, in part, this appears to be due to them being hyper-defensive, in that they tend to overreact to benign contact.[46] Thus, the isolates seem to fail to match their emotional response to circumstances in an appropriately scaled manner. In a beautiful set of studies, a group of Dutch researchers has characterized the emotional inadequacy of rats that have been reared in social isolation.[47] When an adult male rat that has been reared as an isolate during the juvenile period is paired with a large, unfamiliar male, it behaves in a manner that attracts attacks by that male. The failings of such rats become even more evident in another test paradigm, in which groups of young, adult rats that had been reared either in social groups or in isolation were placed in large enclosures containing a colony of adult rats, with one resident male being the dominant animal. The rats reared in isolation were attacked more often by the resident, dominant male than were the rats that had been reared in social groups. The latter decreased their level of activity and moved into a huddle on top of a platform in one corner of the enclosure. The isolates continued to move around the enclosure and failed to take advantage of the sanctuary offered by the platform – and, by doing so, continued to attract attacks from the dominant male. Further, these isolates were also physiologically different to the rats that had been reared as a group.

When we are faced with a stress-inducing situation, stress hormones such as corticosterone are released by our adrenal glands, elevating the circulating levels of these chemicals in the bloodstream. A normal response is to release large quantities of these hormones quickly, as they mobilize the energy systems of the body, preparing the animal to deal with the stress-inducing stimuli – the "flight or fight" response. Then, as the stressful situation subsides, so do the levels of these hormones in the blood. Thus, in such a scenario, a quick elevation of these hormones, followed by a relatively rapid dissipation, is normal. After having been confronted by the resident dominant male, this was the pattern shown by the group-reared rats, but not by the isolates – their corticosterone levels spiked more slowly to a high level, and took longer to subside, so they remained stressed for a longer period.[48]

As the authors of these elegant experiments suggest, play experience may fine-tune the coping skills needed for dealing with different and unexpected social situations. But play may prepare animals for more than just the vicissitudes of social life. When rats that had been isolated from just after weaning (twenty-one days) to around puberty (sixty days) were tested on an elevated plus-maze, some interesting differences were found when their performance was compared with that of their socially-housed counterparts.[49] As the name implies, the elevated plus-maze is a maze that sits on a platform above the ground. The maze itself consists of several, narrow arms, radiating from a center, some of which are enclosed (like a tunnel, leaving the rat hidden) and others that are open (leaving the rat exposed). This clever, yet simple, device has become standard for testing anxiety in small, laboratory rodents. Those rats that are the most anxious explore fewer arms and are less likely to move out into the open ones than those that are less anxious.[50] The performance of the isolates on the maze was impoverished when compared to that of the animals that had been reared socially, which indicated that they found the task more stressful. When the isolates were injected with an anxiolytic – an anxiety-relieving drug – they performed as well as the rats that had been reared socially.

Similarly, when the rats that had been reared socially were injected with an anxiogenic – an anxiety-inducing drug – they performed as poorly as the isolates had done.[51] Again, it would seem that rats reared in isolation during the juvenile period tend to overreact to situations, be they social or non-social.

There is some evidence that it takes a shorter period of social isolation during the juvenile period for social deficits to appear, and longer for non-social deficits to emerge – one to two weeks of isolation as opposed to isolation for the entire juvenile period.[52] There is also some evidence that non-social deficits are more likely to be produced when rats are isolated in some phases of the juvenile period as compared to others.[53] Obviously, more research is required to determine whether there is differential susceptibility to social isolation during the juvenile period for developing coping skills for social and non-social situations. For the present purposes, however, and irrespective of what the deficits are, it seems that social isolation results not in the loss of specific social or cognitive skills, but instead, affects how rats cope with situations that require the application of these skills. The common impairment is an emotional one: by failing to calibrate their emotional response to the situation appropriately, they are unable to use their motor, social, or cognitive skills effectively.[54]

So, what exactly do animals learn during juvenile play fighting that is relevant to developing coping skills? Well, one thing that they learn is that, sometimes in life, things that are enjoyable can also be painful – many times we have heard the thud of one rat falling onto its back when its partner crashes down on top of it. In a situation like this, the play would continue unfettered, but in an aggressive or sexual situation, similar painful stimuli would elicit a very different response. That is, the sensation of pain is contextual: even if the intensity of the pain is identical in two different situations, how to interpret that pain, and so determine what to do about it, depends on its context.[55] Another way to think of the play experience is as a means by which to prepare for the unexpected.[56] Whether these subtle differences in terminology reflect real differences in the processes involved or not, at this stage of our knowledge, it seems reasonable to conclude that play experience in the juvenile period makes rats more resilient and better able to deal with situations. Further, some evidence for this proposition comes from species other than rats.[57] We will return to the comparative literature in a later chapter, but for now, let us continue with the rat's tale so that we can better understand how play may serve this function.

The Uniqueness of the Juvenile Play Experience

Recall that when defending against nape contact, a juvenile rat is most likely to turn completely over to lie on its back. In doing so, the rat moves its nape away from its partner, and is able to block further attempts to gain access to its nape. But there is something odd about using this tactic. By turning over onto its back, the defender limits the options that it has in both defense and offense because, after all, its partner is standing over it and is actively blocking or restraining its movements. Moreover, there are other perfectly good ways for rats to defend themselves during play fighting that do not involve relinquishing much control.

The most obvious alternative is the tactic most commonly used by the rat in the week preceding maturation into a juvenile (approximately eighteen to twenty-five days, when weaning is completed); that of turning partially onto its side, keeping one or both of its hind feet planted on the ground. When employing this tactic, the rat turns to face its partner and moves its nape away, and can then push and restrain its partner with its forepaws, push against it with its flank (a "hip-slam"), or it can rear up and turn completely around to face towards its partner (an "upright defense"). In other words, the defender has several options when using the partial rotation tactic, while still allowing a wrestling-like play fight to occur. Indeed, this alternative, partial rotation tactic is the most common form of defense in many other rodents during the juvenile period.[58] The predilection for juvenile rats to use the complete rotation tactic, rather than the partial rotation tactic, is all the more unexpected given their behavior around the time of weaning.

At this time, young rats are more likely to use the partial rotation tactic even though, when they do attempt to use it, they often overbalance and fall over. Similarly, when they try to eat a piece of food that they are holding between their forepaws, they will gradually fall forward. The basic problem here is that they are not yet able to balance their body weight on their hind feet effectively, which would require them to shift their center of mass backwards.[59] Thus, the weanling rats are employing a defensive tactic during play fighting for which they are ill equipped. By the time they mature into juveniles, rats are able to use the partial rotation tactic competently, but nonetheless, switch to mainly using the complete rotation tactic. During puberty, male rats switch from mainly using the complete rotation tactic back to using the partial rotation tactic.[60] There is no physical reason that we can ascertain as to why weanling rats persist in using a tactic that they have difficulty in using, especially when they can readily use the complete rotation tactic. Nor can we explain why juvenile rats switch to the complete rotation tactic when, by this age, they are perfectly able to execute the partial rotation tactic. Our suspicion is that the juvenile's predilection for the complete rotation tactic is designed to maximize the experiences that play affords the animal for training its emotional responses to match unpredictable situations appropriately. Two other aspects of play fighting in the juvenile period support this notion.

First, when the defending rat rotates to fully supine, the animal standing on top behaves in a peculiar manner. The obvious advantage of being on top is that by using its forepaws, the movements of its supine partner can be restrained and its counterattacks blocked. In order to use its forepaws effectively, and to make any necessary, accompanying shifts of body weight with its upper trunk, the rat stands with its hind feet on the ground; this provides a solid base of support for its upper body movements. In this anchored position, the on-top animal has considerable stability and so can maintain a high degree of control over both its own movements and those of its supine partner (figure 4.5a). However, rats – of all ages – will, at times, do something during their play fighting that seems stupid while on top – they stand on their supine partners with all four of their limbs (figure 4.5b). When they do this, they have less control over their own movements, and, certainly, a reduced ability to restrain their partner's movements. Imagine being on one of those carnival rides that involves trying to remain upright while standing on a wobbling platform – this situation must be akin to what a rat faces when standing on top of a squirming partner. Given that about 80% of the time, the on-top rat stands mainly in the anchored position, it is possible that the rat may occasionally become overly excited and simply make a mistake. But this does not explain what happens in the juvenile period, in which there is a marked increase in the use of the unanchored, on-top position. An on-top juvenile is more likely to stand on its supine partner with all four paws. Furthermore, there is no question that standing on its partner in this way reduces the ability of the on-top animal to remain in control of the situation; a supine rat that has an unanchored partner standing on top of it has a 70% chance of launching a successful counterattack, compared to only 30% when its partner is anchored.[61]

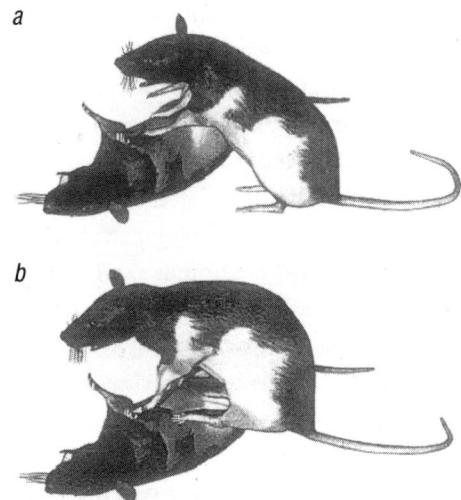

Figure 4.5 When engaged in play fighting, rats often adopt a posture where one animal is standing over the other rat, which is lying on its back. However, the posture of the rat on top can take one of two forms: it can hold its partner with its forepaws while standing on the ground with its hind paws (a) or, it can stand on its partner with all four of its paws. (From Foroud, A. & Pellis, S. M. (2003). The development of "roughness" in the play fighting of rats: A Laban Movement Analysis perspective. *Developmental Psychobiology,* **42**, 35–43. © 2003, Wiley; reprinted with permission from John Wiley & Sons, Inc.)

Because play fighting in rats occurs most frequently in the juvenile phase, it could be argued that juveniles are overly stimulated during this period and so are simply more likely to make such silly mistakes. Not so. Whether engaging in many bouts of attack and defense or just a few, juveniles are more likely to adopt the unanchored position. Decorticated juveniles, at all ages, also use the unanchored position at this same high frequency – this includes periods both when play is sparse and when play is frequent.[62]

The second piece of evidence that indicates that play fighting is designed to calibrate emotional responses to unpredictable situations comes from the involvement of the cortex. The cortex modulates play fighting in such a manner that, in juveniles, there is a high frequency of both the complete rotation tactic and the unanchored on-top position.[63] The absence of any deficits in the rats' ability to initiate and execute playful sequences in the absence of a cortex strongly suggests that this cortical modulation is specifically designed to ensure that juveniles play in the manner that they do. That the age-related modulation in the content of play may be limited to one specific portion of the cortex, the motor cortex, without there being any repercussions on the ability of other areas of the cortex to exert different kinds of modulation over play fighting, further supports the hypothesis that the juvenile-typical pattern of play has evolved to provide specific experiences. Similarly, the finding that of all types of social behavior that can be engaged in, play fighting is the most rewarding for juvenile rats, but not for weanlings or young adults, suggests that juveniles are motivated to experience play.[64] Further, the fact that the juvenile play experience involves a unique pattern of play which is orchestrated by a novel regulatory mechanism involving the motor cortex, makes it less likely that this pattern of play in rats has arisen through some random or correlated change in the organization of the system – even more so given that this cortical regulation does not appear to be present in other murid rodents. Rather, the evidence suggests that the unique pattern of play in juvenile rats is a product of natural selection and therefore serves a particular function.[65] But what might that function be?

Learning to Cope

If we view the behavior of the defender and the attacker in concert, what we see is that, in the juvenile period, both partners are adopting tactics that reduce their control over their own and their partner's movements. In this state of reduced control, the rats experience unpredictability. Such experience may be valuable because the world is often unpredictable, and you have to take that in your stride in order to determine when and how you need to act. A large part of that lesson likely involves learning how to calibrate one's emotional response accordingly. Remember that adult rats that are deprived of play fighting as juveniles overreact even to benign contact, and that this lack of sensitivity to the actions of others makes them vulnerable to attack. But does more play really improve your ability to calibrate your emotional responses appropriately? Growing up without any opportunity to engage in play fighting certainly does seem to lead to problems, but while some play is clearly good, does it necessarily follow that more play is better? There are tit-bits of evidence that suggest there may be an optimal level of play.

In one study from the early 1970s, researchers placed juvenile male rats in one of three different conditions. In the first condition, the so-called "control" condition, male rats were reared over the juvenile period, in standard laboratory cages in social isolation;[66] in the second, pairs of juveniles were reared together, in standard laboratory cages; and, in the third, groups of rats were reared in large cages which offered them the opportunity to climb, explore, and interact with many inanimate objects, as well as with other rats. Then, as young adults, they were tested, in the standard measure of the time, for their performance with a sexually receptive, adult female. As expected, the rats that had been reared in isolation performed poorly when compared to those that had been reared in pairs. But the results for the rats that had been reared in the enriched condition were surprising. They also performed more poorly than the rats that had been reared in pairs! This experiment suggests that too much play during the juvenile period may be as harmful to one's development as too little.[67]

There are also problems if play experience produces rats that are too confident. When rats explore a new area, they maintain contact with vertical surfaces, which gives them some measure of protection from potential predators. Then, as they gain familiarity with the new environment, they make increasingly longer forays, away from vertical surfaces, into the unprotected areas. However, rodents that venture further from protective covering are also those more likely to be taken by predators.[68] Rats and mice that have been reared in social isolation are very reluctant to move away from vertical surfaces, whereas ones that have been reared socially do so more readily. So, it may not be unequivocally good that extensive play experience produces animals that are overly confident and assertive. Too little play may produce animals that overreact to novel encounters with the social and non-social world, but too much play may produce animals that are too unresponsive to the dangers of the world. Consequently, although the evidence seems strong that juvenile play fighting experience is important for the development of a competent adult, it may be that there is an optimal level of play needed to achieve this function. As sages have stated throughout history: all things in moderation.

Irrespective of the amount of play experience that may be needed, the peculiar modulation of the content of play fighting by juvenile rats, which leads to repeated exposure to unpredictability (albeit in a safe setting), suggests that play is used as some form of emotional calibration. How exactly does this work? There exists a large literature on the effects of early postnatal stress on the conditioning of an animal's stress-response system. Contrary to the popular view that early stress may have dire consequences for later development, it seems that the absence of all stress may be worse. For instance, removing a young mouse or rat from its mother, and then subjecting it to some handling treatment, such as stroking or blowing air on it, is stressful, yet rats that have been handled in this manner tend to do better as adults than ones that have not been handled. When adult rats are exposed to a stressful situation, they react by increasing their production of corticosteroids. In handled rats, these hormones dissipate rapidly from their bloodstream, whereas the unhandled rats retain high levels of corticosteroids in their blood for a longer period. Similarly, adult rats that have been handled as infants are bolder in their exploration of novel enclosures, and when old, are better able to learn new tasks, and have a slower age-related degeneration of the brain. In short, rats that have been handled as infants are more resilient as adults. However, the stress cannot be too great. If infants are separated from their mothers for long periods and the stimulation to which they are subjected is very intense, then the outcome is more negative – as adults, the animals over-react to stressful stimuli and are less capable of learning tasks. The beneficial effects of stress experienced early in life seem to occur when that stress is not too great.[69] Again, all things are best in moderation.

The effects of being handled early in infancy appear similar to those we have already described for rats with and without juvenile play fighting experience. This similarity may not be coincidental. These early handling effects appear to engage the mechanisms that are naturally activated by maternal care. During the day-to-day course of mother–infant interactions, the mother rat will lick, step on, and carry her infants in her mouth. She will also leave them periodically so as to feed or explore the surrounding area. Therefore, a normal part of an infant rat's experience is to be both unexpectedly handled and deprived of its mother's presence for brief periods.[70] These early experiences provide the young animal with exposure to some of the trials and tribulations of life, and so condition its stress-response system to a world that is not always uniformly good, possibly doing this by providing a range of acceptable discomfort, rather than a narrow margin. But at this point in the infant rat's life, there are limitations to its ability to make sense of its experiences. Its eyes and ears are closed and, even though its tactile abilities are considerable, making comparisons, such as pairing a particular feeling with particular stimuli, is likely difficult. Thus, in the infant rat, the stimuli that elicit stress are likely to be relatively generic, with only differences in intensity being calibrated.

By the juvenile period, a young rat's sensory systems are all fully functioning and its cognitive capabilities have developed (or are in the process of developing). It therefore has the machinery to make more precise connections between specific instances of loss of control, pain, or emotional distress and particular experiences in the world. That being the case, we suggest that play fighting, especially the juvenile form that enhances the experience of unpredictability, provides a perfect means by which to fine-tune the stress-response system, or, more generally, emotional reactivity, and so produce an adult animal that is capable of subtle and nuanced responses to novel and potentially dangerous situations.[71] Certainly, male adult rats that have not engaged in play fighting as juveniles have a narrower range of options when dealing with novel situations than those that have, even when they have had the normal infantile rearing experience. In addition, while both mice and rats depend on early infantile experiences to condition their stress-response systems, rats engage in play fighting, whereas mice do not, or, at least, when compared to rats, do so in an impoverished manner. It is perhaps not a coincidence that adult rats seem to have a greater range of options than do mice when dealing with both social and non-social situations. Further, rats learn tasks more quickly and to a higher level of proficiency than do mice, and can interact with other rats in more subtle ways than mice in both sexual and non-sexual encounters.[72] That is, play fighting, of the kind fashioned in the juvenile period of rats, provides a supplementary shaping experience for fine-tuning the stress-response system: a system that in all mammals is given its basic form by early mother–infant interactions.[73]

The evidence seems fairly convincing that the play fighting experience of juvenile rats fine-tunes the animal's emotional reactivity and this, in turn, affects how well it can bring to bear its physical and cognitive skills on various situations in adulthood. Indeed, the peculiar pattern of play fighting in the juvenile period makes sense if this experience is a "training ground" for rats to learn how to cope with life's unpredictability. But is there more to the play fighting experience than can be explained by this function?

The Many Benefits Problem Revisited

Not all the deficits that arise from lack of play fighting experience in the juvenile period require the actual physical contact afforded by play fighting. When deprived of play fighting experience as juveniles, rats show a range of cognitive, sexual, and social deficits. As shown by Dorothy Einon and her colleagues, juveniles reared in social isolation grow into cognitively competent adults if they have but one hour per day of exposure to a peer, with whom the majority of the time is spent in play fighting. In the 1950s, the Forgays showed that if rats that had been reared in social isolation as juveniles were given the opportunity to watch normally raised juveniles engage in peer interactions, they would also grow into adults that were not cognitively challenged.[74] Thus, for rats, merely observing other rats play fighting is enough to protect their developing brains from the ill effects of being raised in isolation. But watching other rats engage in normal interactions seems to be insufficient for these rats to develop normal sexual competency, or the ability to execute socially coordinated actions.[75] In guinea pigs that have been reared in social isolation, however, watching other, normally raised, juveniles engage in sexual play is sufficient for them to develop a competent sexual repertoire.[76]

No amount of post-pubertal social experience appears to ameliorate the lack of ability shown by rats deprived of play experience as juveniles to modulate their play, or other social responses, to different social partners correctly.[77] They can, however, be induced to perform sexually in a normal manner if the male is provided with a stationary female. Moreover, after one successful mount with intromission, these rats appear able to modify their behavior with a more mobile female appropriately – an example of "single trial learning!"[78] In contrast, monkeys that have been deprived of experiencing play as juveniles remain impaired in their sexual abilities, even after a successful mount.[79]

At present, the data are simply insufficient to determine whether all these diverse effects across various species can be adequately explained by the role of play fighting experience in fine-tuning emotional reactivity. We suspect that some of the effects of play fighting are likely to involve direct training in the execution of movements, but others are indirect, where it is the modification of the emotional response that results in the execution of the movements being rendered more effective. In some circumstances, the effects of play fighting may work as a transitory scaffold that permits the correct execution of actions. As in the case of the single trial learning of mounting, once the animal is proficient in these actions, no further practice is required. As well as this diversity, there are likely to be varying evolutionary changes, with some steps occurring before others, and not all lineages exhibiting all steps. This issue will be dealt with in depth in chapter 6. At this point in the book, we want to warn readers to be wary of accepting a "one size fits all" explanation for the function of juvenile play fighting. For rats, learning to cope is clearly one benefit gained from play fighting and, given the peculiar pattern of juvenile play fighting, this benefit has probably been an important function that has selected for those playful qualities. But, equally clearly, there are also other benefits that have shaped aspects of the play fighting experience. The ability of a rat to modulate its play fighting with respect to the identity and actions of its social partners is a possible direct benefit derived from the experience of play fighting as juveniles.

Brain Begets Play and Play Begets Brain

Following puberty, male rats that live together develop dominance relationships, with one male becoming dominant and the others subordinate to it. As these relationships are developed, their pattern of play with one another changes. Dominants and subordinates respond to playful attacks by subordinates in the typical, adult male manner, in that they are most likely to use the partial rotation tactic. However, when subordinates are playfully attacked by dominants, they are most likely to revert to the juvenile-typical pattern of mainly using the complete rotation tactic.[80] Ablation of the orbital frontal cortex (OFC) produces subordinate rats that fail to make this discrimination; these rats are most likely to use the partial rotation tactic, irrespective of the social status of their partner. What is more, OFC-damaged rats also fail to adjust their responses depending on the identity of their social partners in non-play social interactions.[81] The OFC thus seems critical in modulating social responses in a way that is sensitive to the identity of the partner. We have already reviewed data that suggest that rats without experience of play fighting as juveniles are similarly deficient in their social competency as adults. There is a linkage.

The prefrontal cortex, of which the OFC is a part, is what is known as the executive area of the brain; a part of the brain that guides one's choice in how to behave in different situations based on the prevailing context, on goals, past experiences, and available alternatives. It is an area of the brain that matures over the juvenile and early pubertal period, and appears to be, during this time, particularly sensitive to experiential feedback. The medial prefrontal cortex (mPFC), another important subdivision of the prefrontal cortex (see figure 3.6), has also been implicated in regulating some features of juvenile play fighting – rats with damage to the mPFC initiate less play and seem to have problems in suppressing the use of defensive tactics that make reciprocity difficult to maintain. That is, these rats are less interested in playing, but once they do play, they play to win. Their lack of reciprocity leads to a greater likelihood of escalating to serious fighting with their intact partner. In addition, imaging studies of humans engaged in social interactions show that both the OFC and the mPFC are engaged in contexts involving cooperation and competition, albeit in complementary ways.[82] Consistent with these studies involving brain damage and imaging, we have also found that play fighting experience during the juvenile period changes the anatomy of the cells in the prefrontal cortex.

Neurons, the main cells of the nervous system that form the communicative circuits that drive behavior, make connections via small points of contact (or near contact), called synapses. These synapses are located on the tips of axons, long extensions from the cell body that make contact with other cells, and on dendrites, shorter and more branchy extensions from the cell body. The major type of neuron in the cortex is the pyramidal cell. These cells have two main dendritic fields, the apical dendrites that branch upward from the top of the cell, and the basilar dendrites that branch laterally downward from the bottom of the cell (figure 4.6). The apical and basilar fields differ in their connections and thus participate in different neural networks. Cells can have more synapses by either having more dentritic branches (i.e. greater arborization), or by having more densely packed synapses per unit of dendrite, or a combination of both. By suitable staining and microscopic visualization, these differences can be counted and compared.[83]

Female weanling rats were housed and reared in one of three groups: (1) with an adult female; (2) with another weanling; or (3) with three other weanlings. Then, at 60 days of age, the animals were sacrificed, their brains extracted and their neurons examined. Since adults rarely play with juveniles, the juveniles paired with an adult female would have had little, if any, opportunity to experience play fighting. The juveniles reared with three other

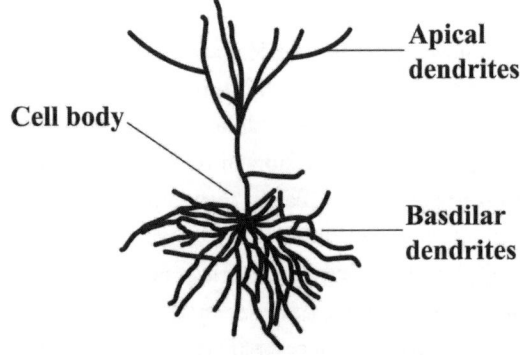

Figure 4.6 A cortical neuron with branching dendrites is shown. This branching, or arborization, is its site of contact with other neurons. Note that there are two main clusters of branches, those emanating from the top of the cell (apical dendrites) and those emanating from the bottom of the cell (basilar dendrites). (Drawing courtesy of Heather Bell.)

juveniles would have experienced more play than the juveniles paired with only one peer. Going from no peer, to one peer, and finally, to three peers, should have provided the juveniles with a progressively incremental level of play fighting experience. We found that play fighting in juvenile rats alters the dendritic branching of the neurons of the prefrontal cortex, with the arborization of the basilar dendrites of the OFC cells increasing and that of the apical dendrites of the mPFC cells decreasing.[84]

Play deprivation leads to a reduction in the ability of the animal to formulate and engage behavioral options dependent on the executive functions of the prefrontal cortex – the same kinds of problems that animals have when they are reared normally but are subjected to experimental damage of their prefrontal cortex.[85] Complementary to these findings is our demonstration of play fighting experience changing the anatomy of neurons in this area of the cortex. Although it is not understood how these changes in neural anatomy affect the functional output of the prefrontal cortex, the fact that they arise through the experience of play fighting during the juvenile period shows that play influences the development of the brain. Even the type of partner to which one is exposed in the juvenile period may be important.

As we have already discussed, the greater amount of play fighting by juvenile male rats compared to juvenile females can be attributed to the effects of exposure to testosterone around the time of birth. Indeed, male rats that are injected with testosterone shortly after birth can be made to play even more. Measures of cortical thickness reveal that these more masculine rats have a greater volume of cortex and that the sex-typical asymmetry of their right hemisphere versus their left one is exaggerated relative to control males that were injected with an equal volume of oil. There is nothing particularly unexpected in this finding. However, when the untreated playmates of the testosterone-treated and oil-treated animals were compared, it was found that the partners of the more playful rats also tended to have greater cortical thickness. Just growing up with a more playful partner influences the development of the cortex![86] Cognitive, emotional, and social skills that are dependent on cortical mechanisms are influenced by play fighting experience in the juvenile period, and this is reflected in changes in the anatomy of the cortex. In turn, changes in the frontal cortex may exert their influences on behavior by regulating subcortical processes. For example, emotional responses are dependent on the activation of the amygdala, and, as we discussed in chapter 3, the amygdala is likely crucial to ensuring playful reciprocity. The prefrontal cortex dampens the activity of the amygdala, which thus prevents emotional overreaction.[87] Here is a potential mechanism by which experience with play fighting in the juvenile period refines the prefrontal circuits, and, in doing so, produces animals that are better able to regulate their subcortically generated emotions.

There is one more piece to this story – there are cortical mechanisms, those involving the motor cortex, that ensure that the form of the play present in the juvenile period is the one that best serves this role of modifying the anatomy, and so presumably the function, of the prefrontal cortex. The existence of a traceable link from the behavior to the brain and then back to the behavior, and the fact that it is organized in this way, cannot be accounted for by happenstance. Rather, it is strong evidence that natural selection has played a hand in fashioning this as a function of juvenile play fighting. While studies of other species have noted that play, of all kinds, may provide such a feedback role in shaping neural circuits at both the cortical and subcortical level, these data are based on correlations between the time of maturation of the brain and the occurrence of particular types of play.[88] In contrast, the data on rats provide direct, experimental evidence that this feedback occurs. Jaak Panksepp and his colleagues have shown that not only does play fighting produce specific neurochemical changes in different areas of the brain, but that it also leads to the production and release of growth factors in the very areas of the cortex in which we have seen anatomical changes.[89] Thus, we could say that *the brain not only shapes play, but that play also shapes the brain.* The full feedback circuit, involving both the cortical and subcortical systems, remains to be charted, but at least we have the beginnings of the circuit at the cortical level.

Before leaving this interplay between play experience and brain development, there is one more caveat we need to consider, as it again highlights the problem of discriminating between the direct and indirect influences of experience on development. Although play fighting experience influences the cell morphology of prefrontal cortex cells, the nature of the experience needed to alter the cells from different regions of the prefrontal cortex may differ. As we have mentioned, during the juvenile period, the amount of play fighting that rats experience can be manipulated by housing them with different partners: a juvenile that is housed with an adult female will experience social contact, grooming and huddling, but little, if any, play. In contrast, a juvenile that is housed with another juvenile will experience the full range of social behavior, including, at least, one hour per day of play fighting, and the amount of play can be increased even more by housing several juveniles together. This increase in play occurs because juveniles are sensitive to a contagion effect; if a juvenile that is little motivated to play is housed in this manner, it will become swept up in the excitement that is created by its more playful partners and will then play as much as a highly motivated partner.[90] Thus, in the presence of multiple partners, an individual rat will play more than when housed with only one other peer. However, when studying the effects of variable play experience on the anatomy of prefrontal cortex neurons, Heather Bell, a graduate student in our laboratory, found a surprising dissociation emerged.

The cells of the mPFc were sensitive to play experience, but there was a threshold effect; to change the mPFC cells, a minimum amount of play fighting experience was necessary, but more play did not add to these changes. Conversely, the cells of the OFC were sensitive to the experience of having multiple partners – having one social partner, be it an adult or a peer, had no effect (figure 4.7). Of course, this experiment confounds two possible factors: in the third group – the one with three peers – the juvenile is experiencing both play and multiple partners. In a subsequent experiment, juveniles were reared with either three adults or three juveniles – there was no difference in

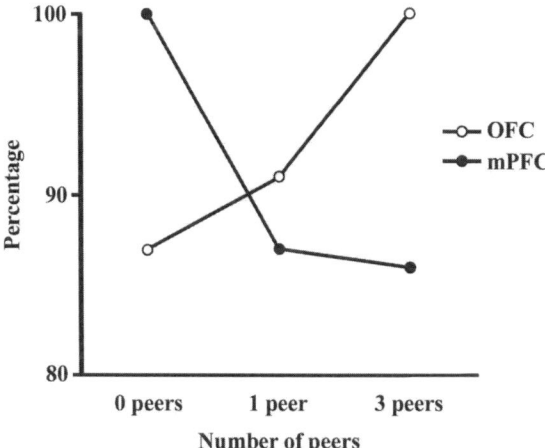

Figure 4.7 The complexity of the dendrites of mPFC and OFC neurons is shown for adult rats, reared, as juveniles, in three conditions. Due to the differences in the magnitude of the branching of the dendrites in the two brain areas, the data are shown as a percentage of the highest value (scoring 100%). Note that the dendrites of the OFC are the most complex when the animals were reared with three other juveniles, and that the neurons of the mPFC decrease in complexity when the animals were housed with at least one juvenile. (Data and graph courtesy of Heather Bell.)

the OFC neurons from the two groups, but the mPFC neurons did differ. Together, these studies show that the cells of the mPFC are responsive to play, irrespective of the number of play partners, whereas the cells of the OFC are responsive to the diversity of partners, whether those partners are playful or not.[91] Therefore, the feedback to the brain which results from play experience may have multiple routes.

For some brain areas, such as the mPFC, the specific content of the play experience may provide the critical feedback, whereas for others, such as the OFC, play may serve merely as a vehicle by which to gain additional, relevant experiences. As we have seen, juvenile rats do not readily discriminate between play partners. For juveniles, any other juvenile interested in playing is an acceptable play partner. In a natural colony of rats, a juvenile rat would not only have siblings of both sexes with which to play, but would also have the juvenile offspring of other females, some slightly younger, some slightly older, and some the same age. Hence, play fighting could serve as a major instrument for a juvenile rat to experience vigorous interactions with a wide diversity of other juveniles. Thus, social competence, which is compromised by being reared in social isolation during the juvenile period, may be thought of as being influenced by two routes during development – *one, sensitive to the experience of multiple partners, and the other, to the experience of play.* The experiences garnered by the animal from interacting with multiple partners may be necessary for them to take their partner's identity into account when deciding how to respond, whereas the experiences garnered from playful interactions may be necessary for the animal to take into account their partner's actions (figure 4.8). As we all know, when confronted with a tricky social

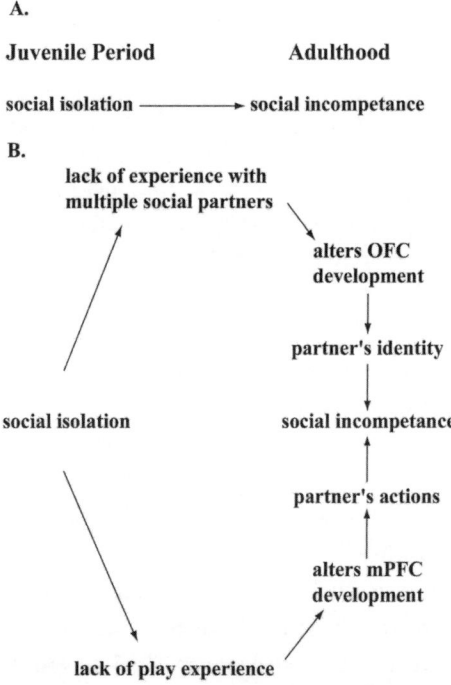

Figure 4.8 Raising juvenile rats in social isolation compromises their social competence as adults (A). However, our work suggests that there are two ways in which social competence can be compromised by social isolation (B). In one route, the experience of playing provides the juvenile with knowledge of how to assess another's actions. This effect occurs via changes in the mPFC. The other route involves gaining experience about differences in behavior across different partners. This effect occurs via the OFC.

situation and deciding how to respond, we have to weigh both the actor's identity and their actions. It remains to be determined what specific experiences affect specific neural systems, how those experiences alter neural anatomy, and, in turn, how this changes the functional capabilities of those neural circuits. Nonetheless, the story to date should alert us to both the direct and indirect ways in which play fighting in the juvenile period can lead to altered brain anatomy and function.

5

PLAYING FOR THE PRESENT

Engaging in play fighting in the juvenile period has repercussions on subsequent development and so on how individuals perform as adults. Findings from rats and some other species leave this conclusion in little doubt. What remains to be understood are the mechanisms by which juvenile play experience influences adult performance. The data that we have reviewed show that this may occur in several ways – play experience may directly enhance motor, social, or cognitive skills, or, alternatively, it may help to produce a better balanced individual emotionally. Such an animal is less likely to be overwhelmed by unexpected events, and so is more likely to bring to bear its motor, social, and cognitive skills on any given situation more effectively. As we will explore in the last chapter, these alternatives may work in concert rather than in opposition. What suffices for now is that there is good evidence that the experience of play fighting in the juvenile period has delayed benefits. In this chapter, we turn our attention to the kinds of immediate benefits that may arise from play.

I Can't Wait until I have Grown Up to Get Value out of Playing

Can it be possible that play makes for a better juvenile? We have already noted that rats continue playing as adults, although at a reduced level, but surely, adult rats cannot be playing so as to train for some future benefit? For rats, at least two immediate benefits of adult play have been revealed. As one of the benefits is common to both juveniles and adults, we will explore that one first. Remember the experiment in which young adult male rats that had been reared either in isolation or in social groups were placed in the territory of a colony of rats, and the dominant, resident male attacked the intruders? With respect to the delayed benefits of play, we noted that this experiment showed that even though the rats that had been reared in groups found ways to minimize the assaults on them from the dominant male, the rats that had been reared in isolation did not. An additional step taken in this experiment highlights an immediate benefit of play. When the dominant male was removed from the colony, the rats that had been reared as a group, but not the rats that had been reared in isolation, increased their level of play fighting and social grooming.[1] For both adults and juveniles, the opportunity to engage in play fighting seems to dissipate their levels of circulating corticosteroids more quickly. Thus, when these levels spike due to a stressful stimulus, the rats that had been reared socially were able to resume other, functionally relevant behavior more quickly.[2] However, if the stress-inducing situation is very intense, or the stimuli inducing the stress are ones that indicate persistent danger – such as the odor of a predator – then play is suppressed.[3] This indicates that when the stress induced is moderate, or when there is no indication of immediate danger, play fighting, like social grooming, appears to be a powerful means of reducing stress levels.

Of course, this all seems very counterintuitive, since from the outset we stipulated that play typically occurs in non-stressful situations.[4] Yet here we are arguing that not only does play occur in stressful situations, but that it may also be instrumental in helping reduce the severity of the stress response! Although being socially isolated is stressful for rats, short periods of social isolation (from a few hours to several days) are an effective way of increasing the amount of play fighting when placed back with a partner. But is it that a certain level of stress is, in fact, that which promotes the occurrence of play? An intriguing experiment by Cheryl Arelis, who was a graduate student in our laboratory, suggests so. Juvenile rats were housed in pairs, and then, a few minutes before being placed in the test enclosure, they were injected with either ACTH (adrenocorticotrophic hormone) or with a comparable volume of physiological saline. ACTH is a stress-related hormone that is released in the brain, enters the circulatory system, and when it reaches the adrenal glands (which sit just on top of the kidneys), stimulates those glands to release corticosterone (in rats) or cortisol (in humans). The initial effect of the release of ACTH is to increase the animal's anxiety, making it less prone to take action. Yet, when the juvenile rats were tested shortly after they had been injected with ACTH, their play increased two to three times more than that of the saline-injected controls. That is, the rats that had been living as pairs until they were tested, elevated their level of play fighting to the same level as that of rats that been socially isolated for 24 hours prior to testing (figure 5.1). Moderate levels of stress or anxiety, then, seem to facilitate the performance of play, even though more severe levels suppress it.[5]

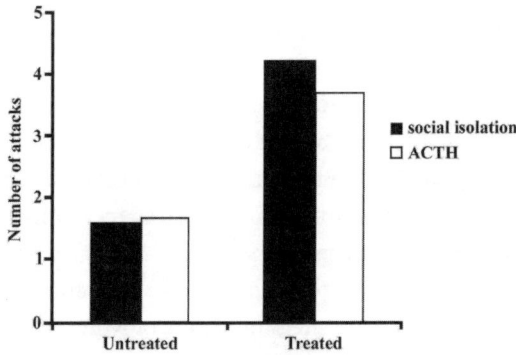

Figure 5.1 Bar graphs show the frequency of launching attacks in pairs of rats (number per minute per pair) that were housed socially and then transferred to a testing cage as pairs (untreated groups). These values are compared to the frequency of playful attacks in pairs of rats that were socially isolated for 24 hours before being transferred to the testing cage and in rats that were housed socially, but were given an injection of ACTH before being transferred to the testing cage (treated groups). Note that for the ACTH comparison, the untreated group was given injections of physiological saline. Both social isolation and ACTH treatment produce more than a two-fold increase in the frequency of playful attacks. Importantly, there is no difference between these two treatment groups. The data on the ACTH treatment are from Arelis, C. (2006) and the data on social isolation are from unpublished records in our laboratory.

Both the experimental literature on rats and the comparative literature on various animals, including primates and carnivores, indicate that play fighting, like other forms of positively reinforcing social contact, serves as a means of social cohesion by reducing tension and stress. For example, Elisabetta Palagi has shown that in captive bonobos (also known as pygmy chimpanzees), in the time-period immediately before feeding, there are elevated levels of play fighting among adults and between adults and immature animals.

This play appears to enhance the adults' tolerance for maintaining proximity with other members of the troop during feeding. Similarly, allogrooming has also been shown to reduce tension and so can have an important role in social cohesion.[6] Both social play and social grooming are sensitive to manipulation of the opioid system, which encompasses part of the pleasure system of the brain, and so, positive social contact, via play or grooming, may regulate stress by activating the release of opioids, which then results in a decrease in anxiety and stress.[7] By using play fighting as part of the behavioral toolkit for regulating the stress-response system, rats, be they juveniles or adults, can quickly adjust to an unpleasant experience or return to normal from a past, unpleasant experience, and so be able to deal with any other unpredictable events that may befall them.

An intriguing study by Adam Miklosi and his colleagues underscores the role of play in the reduction of stress. Hungarian police and border guards have units that use dogs – German shepherds – for their patrols. Policemen and border guards were asked to play with their dogs for a few minutes, and the dogs' salivary levels of the stress hormone cortisol was measured before and after. Bizarrely, in the policemen's dogs, the cortisol level increased, but in the border guards' dogs, it decreased. On closer inspection, the researchers found that, yes, both the policemen and the border guards played with their dogs, but they did so in strikingly different ways. Whereas the police commanded their dogs to play, the border guards engaged their dogs in play by grabbing, tussling, and petting them. As we noted in the very first chapter where we dealt with the definition of play, an important component of play is that it is voluntary. Being ordered to play is apparently not conducive to eliciting the positive emotions normally associated with play, and so may induce stress. The positive emotions elicited by friendly social contact lead to the "desire" to play, which, in turn, produces the stress-reducing effects.[8]

As yet, we do not fully understand how friendly, social contact facilitates stress reduction, and, in particular, what the special features are that may be present in play that perform this function. The answer to this may lie, in part, in the neural circuits and the chemical signals involved in social bonding. These neural systems have been intensively studied in monogamous species of mammals that form intense social bonds. Central to this bonding are the roles of two neurochemicals – oxytocin and vasopressin – chemicals that, when experimentally manipulated, can alter the bonds between pair mates. There are preliminary data that suggest that these neurochemicals can also influence the expression of play. What is clear from this work is that the action of these neurochemicals makes being close to one's partner rewarding, and being separated, aversive. The causal circuits by which these systems operate and are linked to the soothing effects of social play and social grooming remain to be determined, but these are certainly avenues worthy of further study.[9]

Play fighting, then, may be one of the tricks that are available to animals to self-regulate their emotional state. Thus, one of the immediate benefits of play fighting available to rats is used at all ages. As seen in chapter 4, play in the juvenile period is often viewed as a means of producing a better adult, but in this case, we can see that sometimes playing as a juvenile helps you to be a more competent juvenile, rather than a better adult. The other immediate benefit that we have uncovered for play fighting in rats is restricted to adulthood.

Play Fighting as Social Assessment and Social Manipulation

In most of the murid rodents that we have examined, play fighting simulates courtship behavior and so, in adulthood, play-like interactions are only seen in sexual encounters. This is true for the Syrian golden hamster, which, like the rat, has a form of play fighting that is distinctly organized compared to the courtship behavior it simulates. Rats are distinctive in that they continue, as adults, to perform a mixture of juvenile-like and adult-like play fighting in non-sexual interactions.[10] The uniqueness of the rat's pattern is evident when males that are unfamiliar with one another are placed in a novel test enclosure – this needs to be a neutral arena, otherwise the resident will treat the stranger as an intruder and will invariably attack it.[11] In a neutral arena, the animals will initially explore the enclosure to gain their bearings, and will then turn their attention to the other animal, at first sniffing them, especially around their anogenital area (figure 5.2). This social investigation serves to identify the other animal, as olfaction is the primary means by which rodents obtain information about another individual, most notably, its sex and status. Even after short bouts of separation, rodents need to remind themselves as to

Figure 5.2 Two rats are shown engaging in anogenital sniffing.

whether they know a particular individual.[12] After repeated bouts of such sniffing, the animals establish that they do not know each other and that the stranger is an adult male. What to do next? Ah, this is where the rat reveals its versatility as compared to other rodents.

Whether these encounters are between house mice, a species with rudimentary play fighting, or between Syrian golden hamsters, a species with complex play fighting, the options available after the other animal is deemed a stranger, are remarkably similar. One or both animals will avoid each other, one or both will launch a serious attack, and, if it is the latter option that is chosen, one of the pair will eventually act in a submissive manner or attempt to flee. The options open to them are to fight or flee, and, if they fight, then this may persist until one animal has conceded victory to the other. Rats, however, have a third option open to them.

After the pair has been introduced into the neutral arena and the preliminaries are over, they may begin to engage in play fighting. The play fighting in which they engage employs the adult male form, which is rougher than the juvenile form. The play fighting may be prolonged, and, in most cases, will lead to one of the rats behaving in a submissive manner, which indicates that it has acknowledged that its opponent is the dominant male in the enclosure. Occasionally, in such a situation, a pair of male rats will escalate the playful encounter into a serious fight, which, as you may remember, can be distinguished by a switch in the target at which the animals are aiming, from nuzzling the nape to biting the rump. The serious fight that ensues will then establish which one of the pair is to be the dominant.[13] Unlike mice and hamsters, rats are able to use play fighting as an intermediate step in establishing dominance, which, if successful, removes the need to escalate to serious fighting. Mice and hamsters have no option but to concede defeat or fight on.

Play fighting is also used within a colony of animals familiar with one another. In a colony of adult rats, the dominant male is the focus of attention of all others; both males and females will approach him and engage him in friendly, social behavior, such as social grooming and play fighting. If any of the adults of the colony playfully attack this dominant male and he then counterattacks, they will respond by using the juvenile-typical defense patterns. Even females, which use the complete rotation tactic as adults anyway, will increase the likelihood of its use when playing with a male, especially if it is the dominant male. So, it seems that the adult rats are being obsequious, in that they will maintain familiarity with the dominant male so as to avoid being forgotten and then subsequently attacked as strangers.[14] For adult males, being on good terms with the dominant male is particularly critical, since whether their presence is tolerated or not affects their access to food and, even, if only occasionally, to a mating opportunity.[15] Thus, play fighting is one of the means by which subordinates in a rat colony maintain a friendly relationship with the dominant male. But there is also evidence for it having a slightly different use – as a means of testing dominance relationships.

In a colony of rats, not all subordinate males are equal – some seem satisfied to remain subordinates, getting the few crumbs allowed them by the dominant, but some are dominant wannabes, and play fighting gives these hopefuls a way in which to try to change the *status quo*. These wannabes can be identified, when subordinates, by their behavior. In one study, we established a colony of three males, allowed them to grow up together and then, once they were young adults, determined which was dominant. Sure enough, one male received most of the playful attacks from the other two, and although the dominant mostly defended itself with the partial rotation tactic, when he playfully attacked the other two, they were most likely to use the complete rotation tactic and end up supine. These two subordinates played less with one another than they did with the dominant, and, when they did so, they were most likely to use the partial rotation tactic. However, closer inspection of the two subordinates revealed that there were subtle differences between them: one initiated a little less play with the dominant and used the complete rotation tactic a little less often. This same individual also tended to play more roughly with the other subordinate. Once they established each other's identity and how they each behaved, we removed the dominant male. After a couple of weeks, the two remaining rats established a dominance relationship, and guess who became the dominant animal? The subordinate rat that when in the triad had been a little rougher in its play fighting became the dominant male.[16]

Play fighting offers the reluctant subordinate a means by which to test its ongoing relationship with the dominant. The subordinate can escalate the play fight into its rougher form by using the partial rotation tactic more often, and, if the dominant tolerates this escalation, it can be ratcheted up, until either the dominant roughly subdues the subordinate or the status relationship is reversed. This within-colony use of play fighting resembles that which we described above for interactions between adult males that are unfamiliar with one another.[17] Therefore, among adult male rats, play fighting can be employed as a means of friendship maintenance and of testing existing relationships with an eye to changing them. This can be stated in a more general form: play fighting by post-pubescent animals is a means of "assessing and manipulating" partners and opponents in the rat race.[18] As we will see in a later chapter, there are many common features to play fighting when used in such a manner, irrespective of how the species may have evolved complex play fighting in the first place. But why is this true for rats and not for hamsters or mice?

For the reasons that we discussed in chapter 1, this question is not easy to answer. The problem is that the dozen or so species of murid rodents for which we have sufficient data only represent a small percentage of all rodents. In the remaining two thousand species of rodents that have yet to be studied, there may well be other species that have retained play in adulthood and have modified it for use in manipulating social relationships. If there were, these cases could then be documented and we could see what they all may have in common. With a data set like this, hypotheses could be postulated and then tested. However, as no such data set exists, this is merely wishful thinking. But all is not lost – there are sufficient data on another group of mammals, a group with no more than 250 species – the primates.

Wisdom in the Trees

Broad surveys of play fighting in animals suggest that the retention of play in adulthood is relatively rare, and that if it occurs, it most likely appears in courtship behavior.[19] Playful contact during courtship appears to help over-come the antagonism between animals that are not familiar to each other. If this were so, then species with the least social contact between prospective mates should be the ones that are the most likely to use playful contact as part of their precopulatory courtship. Scouring the literature, we identified data on thirty-six species of primates from all families and sub-families for which reasonable inferences could be made about their social system and the pres-ence of play in courtship. The social systems were analyzed as to whether they afforded prospective mates the opportunity to associate outside the mating season. Species were rated with high, low or moderate levels of male–female association. As play appeared to vary both in the frequency of its use and in its complexity, a prevalence score ranging from absent to highly prevalent was assigned to each species. Using Independent Contrasts to control for related-ness, a correlation revealed that there was a significant negative relationship, with the prevalence of play decreasing as the strength of male–female associ-ation increased. Indeed, 40% of the variation in the prevalence of play was accounted for by the degree of non-sexual social association. It thus seems likely that the need for playful contact in courtship is an important factor in determining whether a species retains play fighting in adulthood.[20]

Two problems contained our excitement over these results. First, what about those cases where play is prevalent in species that have high rates of male–female association? Second, as rats use play fighting for assessing and manipulating non-sexual social relationships, this finding in primates has lit-tle value as an explanation for the retention of play in adult rats. A re-exami-nation of the primate data provides two helpful findings. If play in adulthood is retained for its use in courtship, then it is likely that its use in non-sexual encounters is a by-product of its courtship function. If that is the case, then the presence of sexual play should account for most of the presence of play in non-sexual contexts. Using Independent Contrasts, non-sexual play was compared to sexual play, and although significant, the former only accounted for 15% of the variance of the latter. More strikingly, social systems that led to infrequent and unpredictable contact between adults of either sex accounted for over 40% of the prevalence of non-sexual social play. When comparing sexual play and non-sexual play in primates using cladograms, non-sexual play was shown to be more prevalent and more likely to be present in a species than is sexual play. Further, non-sexual social play is more likely to be present

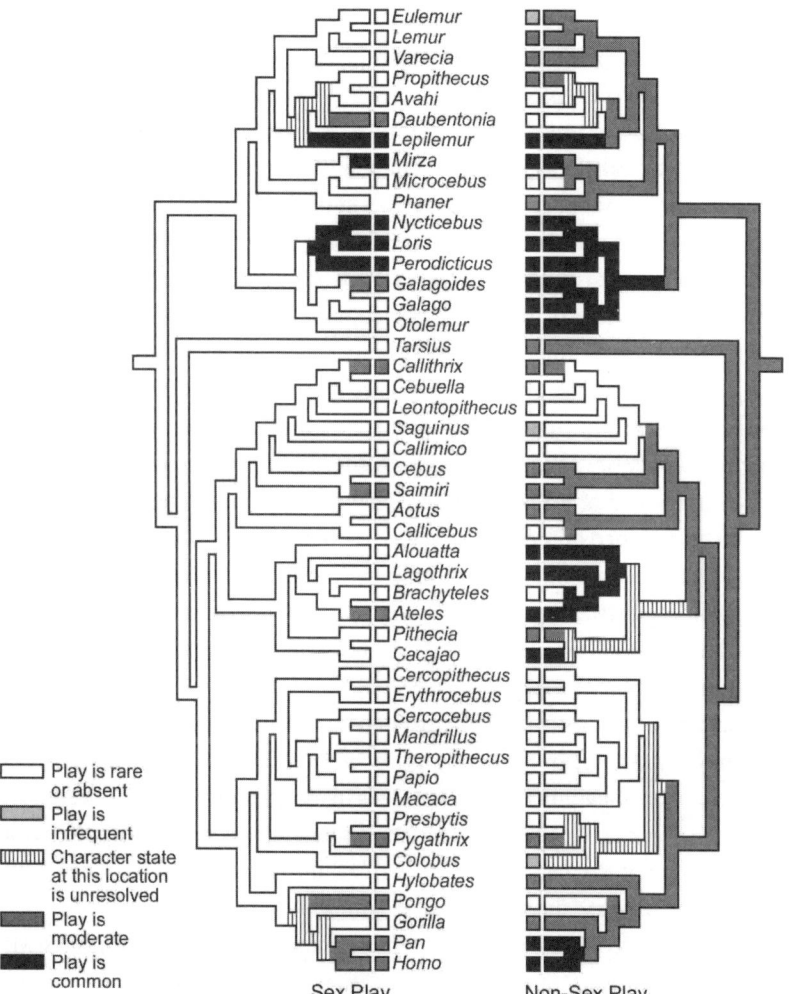

Figure 5.3 The occurrence of play fighting used for sexual and non-sexual purposes in adult primates is shown mapped on cladograms. The names represent the scientific name of the genera. The large clade at the top of the page (from *Eulemur* to *Otolemur*) represents the strepsirrhirines, and the clade from *Tarsius* to *Homo* represents the haplorrhines. The New World monkeys incorporate the animals from *Callithrix* to *Cacajao*, the Old World monkeys from *Cercopithecus* to *Colobus*, and the apes and humans from *Hylobates* to *Homo*. (From Pellis, S. M. & Iwaniuk, A. N. (2000). Adult-adult play in primates: Comparative analyses of its origin, distribution and evolution. *Ethology*, **106**, 1083–1104. © 2000, Blackwell, reprinted with permission.)

in ancestral groups (figure 5.3). Thus, contrary to the initial expectation, non-sexual social play is more likely to have given rise to species which then co-opted its use in courtship.[21]

The findings on rats, together with the analyses of adult play in primates and the data from other species (see chapter 7 for further elaboration), all support the conclusion that play fighting, when it is retained in adulthood, is used as a means of social assessment and manipulation. In general, our impression is that in species for which social interactions among members of a social group are infrequent or otherwise variable, play fighting is often retained in adulthood as a tool for social assessment and manipulation. For example, even though bonobos and chimpanzees live in large social groups, these groups can splinter into subgroups that then wander off on their own. Thus, individuals may only sporadically encounter one another, which requires re-acquaintance and re-assessment. Not surprisingly, these species have among the most developed adult–adult pattern of play. Rats live in intricate colonies while Syrian golden hamsters lead solitary lives; this may be part of the explanation for why rats, but not hamsters, engage in adult–adult play. What about mice? They live in colonies, yet they do not play as adults, and, at best, their play is rudimentary.[22] Both the cross-species distribution of adult–adult play and the variation in its complexity need to be compared to make sense of these patterns fully. Again, given the greater knowledge about social systems in primates, they are likely better placed to provide the answers. Unfortunately, even with the recent studies by Elisebetta Palagi and her colleagues, which provide some of the necessary details of adult play in primates,[23] given that 50% or so of primate species engage in adult–adult play, there are still too few detailed studies available to provide more than a tantalizing glimpse of what may be possible.

Delayed and Immediate Functions of Play Revisited

In the early 1960s, one of the first papers to describe play fighting in rats was published. Now, some 40 years later, due to the efforts of many researchers around the world, we are developing a detailed understanding of this behavior in rats, its development, sex differences, neural control mechanisms, and the functions that play fighting offers.[24] We have discovered that play does not have one particular function, but that it has multiple functions, each one involving some behavioral reorganization that makes it highly likely that natural selection has acted so as to ensure that the appropriate consequences ensue. What is most interesting about these functions is that there are *both* immediate and delayed ones. In the juvenile period, experience in play fighting makes for adults that are better able to deal with the vicissitudes of life – a classic delayed benefit. What remains to be determined is whether all the improvements in coping arise from a form of emotional fine-tuning or whether, in addition, there are specific improvements in cognitive, motor, and social skills that are independent of the general improvement in coping skills. The rat model that we now have for studying play fighting is sufficiently well developed to enable very precise predictions to be made and tested, if not by us, then by the next generation of researchers.

Two distinct, immediate functions of play fighting have also been characterized. At all ages, play fighting is a behavioral tool for rats to regulate their stress-response system. Such play may be one way of delivering endogenously produced opioids in a form of self-medication. Again, the precise mechanisms remain to be characterized, but it has become clear that when stressed, rats can use play as a means by which to calm themselves. The other immediate function applies only to adults, or more generally, to post-pubescent animals, that can use play fighting as a tool to assess and manipulate social partners. So, we have one delayed function that is idiosyncratic to juvenile rats, one immediate function that is idiosyncratic to post-pubescent rats, and one immediate function available to rats of all ages. At the very least, the rat's tale should warn readers to be skeptical of statements like "THE function of play is to…" We should also note that it was only by contrasting the play behavior of rats with that of many other close and distant relatives that the peculiarities of the play in rats became evident and so could point the way to likely functions. But, in the end, a function arises from the beneficial effects of engaging in some action. Given that many other rodents do not have the functions for play fighting that are present in rats, how did rats co-opt rodent-typical behavior into novel functions? To answer this question, we need to bring together what we have learned about the mechanisms responsible for organizing and regulating play, their comparative distribution, and the particular consequences that bring benefits.[25]

6

TAKING SPECIES DIVERSITY
SERIOUSLY

After having read chapters 3, 4, and 5, you should now understand our earlier reluctance to be overly zealous with both the definition of play and its function. Just by looking closely at a limited range of mammals, the murid rodents, and a small slice of this rather numerous group at that, it is evident that even for one form of play – play fighting – there are some species that play and some that do not, and of those that do, they do so with varying degrees of complexity. How can the play fighting of a house mouse that rarely engages in playful attack, and, when it does, limits its defense to evasion, be compared to that of a rat with its frequent play and complex maneuvers? Simply claiming that "play" is play and that it has one or more functions does an injustice to the diversity of play present in the natural world. To feel satisfied that we understand nature requires that we can explain that diversity.[1]

Although broad definitions that can encompass many phenomena into one category are sometimes useful, they run the risk of pigeonholing behaviors that only really make sense when species are compared within a clade of related species.[2] For example, in figure 3.9, applying a definition of play would yield two states, those species that play and those that do not. It is only by taking into account the variation in the complexity of the play of the species that play, that clues can be found for how play may have evolved, and, most critically, for identifying the neural changes that underlie that evolution. As we have seen in chapters 3 and 4, identifying the neural changes that underlie the evolution of play provides a concrete basis for testing hypotheses both about the mechanisms that generate play and the processes that may have led to its evolution. In this chapter, we will integrate mechanism and function, and in doing so, posit a model that can account for the diversity of play fighting in murid rodents. This model will also give clues as to how the underlying mechanisms can be modified to produce patterns of play not seen in rodents, but are evident in other mammalian groups, such as primates, including humans.

Mouse Play to Rat Play Revisited

Because evolution involves transforming existing forms into new forms, what are the most likely building blocks that represent the precursors of play? One place to begin this search is with how behavior develops. Most often, behavioral systems, such as sex or aggression, develop in a piecemeal fashion. That is, over some developmental time period, components of the behavioral system emerge and are gradually integrated. An example of a finely studied behavioral system is dust bathing in the chicken.[3] This involves a series of behavior patterns that move dust particles through the bird's feathers; these actions remove the excess oils and parasites that may twist and misshape them. The dust bathing sequence begins with the chicken scratching and pecking at the substrate. The chicken first drops one wing and then the other, while simultaneously shuffling its feathers. It then rolls on the ground, first onto one side, then the other, and finally squats, with its body feathers raised. More shuffling ensures that the dust released by the pecking, scratching, and wing movements is passed through the feathers to their base. Finally, the chicken stands up and gives its body a good shake to rid it of the excess dust. From hatching, it takes about two weeks for all of these behavioral elements to emerge, and they appear in the chick in the order in which they are performed in the adult in the completed sequence. Thus, scratching the substrate emerges earlier in development than rolling onto the flanks. But there is more to this, or any behavioral system, than going through the relevant actions.

Initially, as these behavioral components emerge, the chicks are indifferent as to whether they are standing on sand or dust or on a wire mesh floor devoid of any usable particles. Whether the chicks are reared in an enclosure with particles (soil, sand, or dust) or in an enclosure with only a wire mesh floor, the order of appearance and the timing of appearance of the behavioral actions remains the same – the behavioral actions develop independently of feedback from the environment. By the time the full behavioral sequence is present in chicks, they show signs that the mechanisms that regulate the occurrence of this behavioral system have emerged. For example, they begin to perform dust bathing with a daily rhythm, and become responsive to deprivation effects. That is, if they are prevented from dust bathing, they will do more of it when they are free to do so. However, these central control mechanisms become active irrespective of environmental feedback – whether dust is available or not. It is only quite late in the developmental sequence that the young birds become sensitive to the perceptual properties of the environment. At this stage, if the chicks are reared on a particular substrate, they will begin to show a preference for executing dust bathing on one substrate over another. Although the motor, sensory, and central control components of other behavioral systems may differ in the order in which they appear, they reveal a comparable pattern of piecemeal development, whether those systems are of birds acquiring song, or of young mammals acquiring adult-typical grooming, righting, and walking.[4]

Why should this piecemeal development of behavioral systems be so widespread? One suggestion is that for an animal to develop a behavioral system in its completed form only at the instant at which it is needed may be more costly than if it were to develop the behavior in a more gradual fashion.[5] Times of transition, such as the juvenile period, when the behavioral systems that are suitable for life as an infant are degrading and the behavioral systems that are suitable for adult life are appearing, are likely to be particularly rich in fragmentary behavior, and so are fertile ground in which to find play-like behavior.[6]

A cost-saving developmental program thus leads to the piecemeal maturation of behavioral systems in the juvenile period, and this immaturity leads to the performance, in many murid rodents, of precocious sexual behavior, with the result being that the behavior occurs before its performers are capable of viable reproduction. In our view, this is the beginning of play fighting. Once such precociously expressed behavior is present in the developmental trajectory, it can be co-opted for use in novel functions, and this is achieved by altering the mechanisms that regulate its expression. But we are getting ahead of ourselves. We must first consider two important caveats. Merely performing behavior precociously is insufficient for that behavior to be labeled as play, otherwise, all immature behavior could be labeled as such and that label would consequently be of little worth. If play does emerge from immature behavior, then the problem becomes one of deciding where to draw the line – what is it that separates play from precociousness? Some fairly sophisticated definitions of play provide a good guide for finding this dividing line,[7] but even so, as small modifications before and after this line are likely, the line is not likely to be a sharply drawn one. The further away from this line that we move, the easier it is to differentiate play from immature behavior. Also, we should not assume that the mere presence of piecemeal development necessarily leads to any given intermediary stage providing feedback for the further development of that behavioral system. For example, in many small carnivores, where both the forepaws and the mouth can be used for catching prey, prey capture with the mouth precedes the onset of manual prey capture during development. But is experience with catching prey with the mouth necessary for the subsequent development of catching prey manually? In the development of both domestic cats and short-tailed opossums, oral capture precedes manual capture, yet in both of these cases, adult animals that have never experienced prey (mice or crickets, respectively) are immediately able to capture the prey with their forepaws.[8] Therefore, just because animals perform intermediate forms of behavior, it cannot be concluded that the performance of such behavior serves a necessary developmental function. The same is also true for behavior that is labeled as play.

In chapter 4, we discussed some intriguing findings in sheep, in which the absence of playful interactions in the juvenile period led to impoverished adult sexual performance. This deficiency did not arise early after birth when play is most frequent, but rather, in the weeks just preceding sexual maturity. The functional feedback that the sheep derived from sexual play cannot, therefore, explain why such play is more common at an earlier age! In other examples, the connection between play and sexual performance is even more deceptive. Recall that guinea pigs that are reared in social isolation during the juvenile period perform poorly, sexually, as adults. However, if juvenile guinea pigs are housed so that they can smell and see one another, but not interact with physical contact, their sexual performance as adults is unimpaired.[9] Thus, the behavior patterns may mature without any significant functional consequence to the behavior patterns themselves, or to the behavioral system from which they are derived.[10]

The initial step, in building play, then, is founded on precociously performed behavior, that does not need to serve any function. But how is this precocious behavior then transformed into play? As they have already done so many times in our tale, murid rodents provide a clue to this transformation. As many murid rodents have litters of up to a dozen pups, juveniles experience multiple opportunities to engage in precocious sexual interactions with their peers. Because some murid rodents, such as rats, live in colonies of multiple breeding females, juveniles not only have their own siblings as play partners, but many others as well. It is thus guaranteed that part of the juvenile's "developmental niche" will provide these types of experiences,[11] and will include precocious sexual interactions with other youngsters. Such experience–expectant systems have important repercussions for evolution.[12]

For example, imagine that you and your pet cat are floating on a life raft, survivors of a shipwreck. You have plenty of water, as well as dried meat and biscuits. You share these rations with your pet and days pass, stranded at sea. At some point, you notice that while your pet's fur has maintained its sheen, your hair is dirty and matted and your skin is dry and peeling. All in all, you are beginning to feel bad, but your pet is still in good shape, despite the fact that you are both equally exposed to the elements. Why are you faring so much worse than your pet? An important physiological difference exists between you and your cat. It has all the biochemical machinery needed to manufacture vitamin C, whereas your own machinery is wanting. Unlike your pet, the lack of vitamin C in your meager diet leads you to develop scurvy, the bane of sailors of an earlier era.[13] But why can your cat manufacture essential quantities of vitamin C, and you cannot? A likely explanation is that your primate ancestors lived on a fruit-rich diet – one with a surfeit of vitamin C – whereas the carnivorous ancestors of your cat did not.[14] As it is very costly to an organism to maintain the machinery needed to make it self-sustaining and able to synthesize all its essential nutrients, natural selection prunes any excess fat from the system. If the environment provides the vitamin C necessary for a healthy life, then there is no need for the organism to manufacture its own. Any variants in the population that possess a less efficient, but also less costly, vitamin C making physiology, will be favored. Moreover, if the environment remains stable for a long time, the population will gradually shift its composition to those animals that do not maintain an expensive and unnecessary capability.[15] The downside, of course, is that the organism is now designed to live in an environment in which it expects to obtain all its vitamin C from its diet.

As a behavioral example, consider young rodents that are developing in their normal environment of a nest in a burrow. Encountering inanimate objects, such as nest debris, fecal bolli, and food items, would be a normal and expected part of their developmental environment. If rats and mice are reared in the absence of such objects, and so are unable to interact with them daily, they can be severely impaired, as adults, in their food-handling and nest-building skills.[16] This suggests that the experience of interacting with these inanimate objects is crucial for wiring the brain appropriately. In a similar vein, the guaranteed presence of precocious sexual interactions removes the need for environment-independent wiring of the neural circuits that are needed to use the behavior effectively. Most critically, precocious sexual experience appears to be essential in being able to calibrate one's own movements to those of one's partner. We have already seen an example of this kind of experience-dependent calibration in the food-robbing and dodging behavior of rats. Rats that did not have peer–peer playful interactions could still produce species-typical dodges of the appropriate magnitude, but could not modify them so as to achieve the correct bodily orientation that took their partners' movements into account (see chapter 4).

With the argument developed thus far, we have established the groundwork for the evolution of the first two stages of play. In the first stage, play-like behavior, or *incipient play*, arises from the immature behavior that appears, during development, in a fragmented manner, as part of a cost-saving program (table 6.1). In the second stage, the presence of this fragmented, immature behavior becomes a reliable part of the experiential world in which the animal develops, and so can substitute for maturational processes that are otherwise insensitive to experience. This second stage may not involve any modifications to the content of the immature behavior expressed, but nonetheless, it provides essential, experiential feedback for wiring the brain, and so, at least functionally, may be thought of as *rudimentary play fighting* rather than simply as immature behavior.

Even though rudimentary play may not involve any organizational changes to its behavior, the frequency of its occurrence may be increased. An increase in frequency makes it more likely that the important experiences gained from performing this behavior are encountered. One way to determine the

Table 6.1

Hypothetical stages in the transformation of immature sexual behavior into play fighting in murid rodents. A similar process could lead to the evolution of play fighting from immature aggression and predatory behavior (From Pellis, S. M. (1993). Sex and the evolution of play fighting: A review and a model based on the behavior of muroid rodents. *The Journal of Play Theory & Research*, **1**, 56–77. © 1993 by Sagamore Publishing Inc., adapted with permission).

Classification	Behavior and context	Consequences and functions
Components of sexual behavior, especially precopulatory elements. These are expressed in a precocious manner during the juvenile period.	None.	*Incipient play* (i.e., play-like behavior).
Under appropriate conditions, precocious sexual behavior becomes a frequent and expected component of the juvenile period. However, at this stage, the juvenile interactions differ little from adult sexual interactions.	Precopulatory behavior becomes necessary for the maturation of normal, adult sexual performance.	*Rudimentary play fighting.*
Some components of precopulatory sexual behavior are elaborated during the juvenile period, making the interactions more clearly different to adult sexual behavior.	These elaborations ensure that the necessary sexual skills are acquired during the juvenile period.	*True play fighting.*
Under some conditions, this sexually derived pattern of play fighting is modified to a more exaggerated degree and co-opted into use in novel domains of juvenile and adult life.	These patterns of interactions now serve non-sexual as well as sexual functions.	*Emancipated play fighting.*

transition from precocious behavior to that which may be thought of as a rudimentary play, then, is by the frequency with which it is performed. Again, extremes are easy to identify. For example, in studies that we conducted on the development of two free-living species of birds, there was a striking difference. In the first four weeks after hatching, young goslings of the Cape Barren goose (an indigenous Australian goose) engage in competitive peer–peer interactions during only 1% of their waking time. These interactions mostly involve quasi-aggressive encounters that follow when the goslings accidentally bump into one another, or, even more rarely, when they run at one another. The bouts are widely spaced and short in duration. Moreover, once the bouts are terminated, the goslings move apart. In contrast, fledglings of the Australian magpie engage in competitive peer–peer social interactions for up to 20% of their waking time. Furthermore, these bouts are prolonged, and on termination, the birds remain in close proximity or quickly resume the encounter – all the hallmarks of play fighting.[17] Admittedly, the peer–peer interactions of the juvenile magpies are both quantitatively and qualitatively different from those of the goslings. Rodents will help us identify the gradations in between.

As we have already discussed, guinea pigs do not physically need to rehearse precopulatory behavior patterns during play-like, peer–peer interactions in order for them to develop normal sexual behavior. They just need to see and smell other guinea pigs as juveniles. In addition, when raised with peers, these animals only engage in play-like interactions for about 1% of their waking time. Similarly, among murid rodents, the time spent in such interactions varies from none (e.g. hopping mice), to up to 17% (e.g. rats), and various values in between (e.g. 1% for Mongolian gerbils, 10% for prairie voles). Like Australian magpies, rats exhibit qualitative changes in their play fighting that distinguish it from sexual behavior. In contrast, in species like prairie voles, the peer–peer interactions are only distinguishable from adult sexual encounters because they do not terminate with mounting attempts or actual copulation. What distinguishes the peer–peer interactions of voles from those of Mongolian gerbils is that voles engage in more of them. The time spent in these precocious sexual interactions is one marker that separates immature sexual behavior from a rudimentary form of play.

Although the time that juvenile animals spend in precocious sexual interactions provides a clue as to when to call such behavior play, a problem remains: how much time do they need to spend in these interactions for us to label it as play? The consequences that these juvenile interactions have on later adult behavior may provide the answer to this question. Juvenile rats that are deprived of peer–peer interactions exhibit a variety of cognitive deficits as adults. However, if other rodents, such as house mice and Mongolian gerbils, that have low frequencies of these peer–peer interactions, are similarly deprived, they do not exhibit these cognitive deficits. Therefore, we could argue that species that have frequent peer–peer interactions have transformed immature behavior into a valuable developmental experience, justifying us in labeling such interactions as play. Although we believe that this conclusion is correct, the evidence is not clear-cut. Juvenile Syrian hamsters, like rats, engage in frequent peer–peer interactions. Yet, if they are deprived of these interactions, they do not exhibit cognitive deficits as adults. Thus, Syrian hamsters appear to be more like mice, even though their juvenile peer-interactions are as frequent as in rats. Why? Maybe cognitive deficits are not the deficits that should be compared. For rats, deprivation of juvenile play leads to both cognitive and sexual deficits. Unfortunately, comparable studies have not been done on rodents such as hamsters and voles to determine if they also exhibit sexual deficits as adults.[18] Our theory is that immature behavior produces a low frequency of peer–peer interactions in the juvenile period. Some species increase the frequency of these interactions above this baseline, and it is this elevated frequency that can be labeled as play. Moreover, this play has consequences for the behavioral system from which those interactions are derived. For example, in murid rodents, this should mean that this play affects the development of sexual behavior, and, for rats, this is true. A test of our theory would be to examine the effects of depriving juveniles in other species that have elevated frequencies of peer–peer interactions. If our theory is correct, then it should be the case that such deprivation in Syrian hamsters and prairie voles will lead to deficits in their sexual performance. Cognitive deficits seem to be unique to rats and these may be accounted for by the further modifications of their juvenile play that will be discussed below.

To reiterate, when elements of sexual behavior are performed precociously in the juvenile period, but at a low frequency and in the absence of any modifications idiosyncratic to that period, then this behavior should be relegated to the status of immature behavior, and not labeled as play. Such behavior may be deemed developmental noise with no adaptive value. When these precocious juvenile interactions persist, they can be co-opted as a means of calibrating sexual movements in a dynamic context and so provide valuable feedback for the development of effective sexual behavior. At this stage, the interactions become functionally valuable, and that value can be enhanced by a simple, motivational change that leads to an increased performance and thus, a greater likelihood that they will serve their developmental function. In neurological terms, this change can be readily affected by a minor alteration to the reward systems of the hypothalamus, nucleus accumbens and amygdala. With a minimal change in the regulation of these juvenile interactions, it becomes reasonable to label them as play, albeit of a rudimentary form (table 6.1).

The third step in the evolution of play fighting is for the juvenile version of adult sexual encounters to be modified, not just in terms of the frequency of performance, but also in the organization of its content. Let's assume that the main beneficial experiences derived from play fighting occur when the animals adopt the supine defense, which leads to prolonged playful wrestles with much bodily contact. In this case, altering the organization of precocious sex so that these experiences are more common during juvenile play would be advantageous. As we move from the second to the third step in the evolution of play fighting, there is a fundamental shift in both behavioral organization and function. For the second step, all that is needed is to increase the frequency of rudimentary juvenile play so as to facilitate sexual development. However, once modifications have been made to the organization of play fighting itself, not only does this behavior facilitate the development of sex, but it also becomes an essential component of the normal developmental experience. With the third step, the playful interactions become both quantitatively and qualitatively different from sexual interactions, and the label of rudimentary play seems insufficient. We suggest that when this stage is reached, the behavior is best labeled *true play fighting* (table 6.1). From a neurological perspective, such changes would likely require modifications of components of the brainstem (e.g. striatum, periaqueductal gray) because, as we saw in chapter 3, decorticate rats still retain these modified patterns of play. But notice that in our discussion of the evolution of play fighting so far, it has still been all about sex. The next step is more radical.

From the standpoint of murid rodents, only rats seem to have moved play beyond its immediate involvement in the development of sexual skills. In rats, the organization of play fighting is more modified than we see in Syrian golden hamsters, and is crucially involved in the development of socially competent behavior that extends into many domains beyond the sexual. Thus, the juvenile version is modified so as to ensure the development of social competence more generally. Even more strikingly, play fighting itself is retained into adulthood as a tool for social assessment and manipulation. To perform these novel uses, the fourth step in the transformation of imma-ture behavior to play fighting involves further modifications in the organiza-tion of the play. At the neurological level, we have shown that these behavioral modifications involve novel control mechanisms emanating from the cortex. Because of the novel organizational changes as well as its expansion beyond its original sexual function, this final form of play fighting can now be consid-ered *emancipated play fighting* (table 6.1).

The steps in the evolutionary scenario outlined above provide a theoretical model or framework that shows how the mechanisms regulating behavior and the functions they serve need to be considered jointly when trying to make sense of the diversity of play in the natural world. Differences in both functions and mechanisms are likely to be present, but these may nevertheless be linked together if natural selection has operated to increase the beneficial outcomes of particular behavioral experiences. In order to achieve this, exist-ing mechanisms would have to be modified.

Our model also cautions us not to assume that all the effects of manipulat-ing juvenile experiences with play are likely to be evidence of functions that have evolved via natural selection. Recall that there can be many benefits accruing from a particular action, but only some of those benefits may actu-ally be evolved functions – others may be incidental by-products. The sequence of stages seen in table 6.1 suggests that in the first three stages, there is progressive improvement in transforming non-functional, precocious sex-ual behavior into a form of play fighting, the experience of which yields improved sexual performance. It is only in the fourth stage of the model that play becomes functionally divorced from sex. Again, this model predicts that if rats are deprived of play fighting in the juvenile period, they should have reduced sexual competence as adults, and they do. It also predicts that such rats would have reduced social competence in non-sexual social situations, and again, they do. But what about the many other cognitive deficiencies that are present in adult rats that have been deprived of play fighting experience as juveniles? Are these general cognitive effects also due to the functions that have evolved for the juvenile play fighting experience?

Just because deprivation in play fighting experience leads to cognitive impairments, it cannot be assumed that the function of the play experience is to facilitate the development of precisely those cognitive abilities. After all, it is not the deprivation of playful performance *per se* that leads to sexual incompetence in guinea pigs – the smell and sight of other guinea pigs is enough to ensure good performance. Similarly, rearing juvenile rats in isolation, but in an environment where they can see (and, presumably, smell) other young rats leads to unimpaired performance on a variety of cognitive tasks.[19] Thus, while play fighting obviously ensures that juveniles will see and smell other rats, the cognitive benefits derived from being in the company of other juveniles cannot explain the organization and frequency of play fighting. The benefits to sexual and social competence, however, do provide us an explanation of why play fighting takes the specific form that it does in this species. As we discussed earlier, there is no magic formula for discovering evolved functions, and experiments are not a privileged source of evidence. Rather, it is the preponderance of evidence, from all sources, which leads to the most convincing models accounting for how function has shaped mechanism and how mechanism has shaped function.[20]

Using this eclectic approach, we have seen that the rich and complex play fighting of rats is essentially a layer cake, with new functions and mechanisms being built on and integrated into the foundations of existing mechanisms and functions. Our model is thus consistent with the general model developed by Gordon Burghardt to account for the multifaceted evolution of all kinds of play throughout the animal kingdom. Primary processes, such as piecemeal behavioral development, sufficient extra nutrition and freedom from predation, make play-like behavior possible, and then, secondary and tertiary processes can transform that play-like behavior into more complex and functionally diverse play.[21] Consistent with both Burghardt's model of play and with our own, which we derived from our empirical analysis of the play fighting of murid rodents, are the behavioral and neural changes that we see in the third and fourth stages of table 6.1. At these stages, juvenile play fighting has acquired its own organizational and temporal features that differentiate it from the maturational processes that give rise to precocious behavior. Therefore, new neural control mechanisms have been added to organize and structure such play during the juvenile period. Indeed, neural control mechanisms that emanate from the cortex and that modulate the brainstem regulatory controls have now been characterized in rats.[22]

Although murid rodents illustrate how these multilayered processes may transform play, there is, however, a problem. Among the sixteen species of murid rodents that we have studied and for which we have sufficient data, there are both gains and losses in the complexity of play fighting (see figure 3.9). What we cannot know is whether the four gradations detailed in table 6.1 have to be gained or lost in sequential order. That is, can a species go directly from stage 1 to stage 4 and vice versa, or, for a species to go to stage 4, does it first have to pass through stages 2 and 3? There are over a thousand species of murid rodents alive today, so using only sixteen to represent this group as a whole means that our cladogram is too coarse to reveal the pattern of evolutionary change – even the seemingly closely related rat and house mouse diverged over fifteen million years ago. Over that time period, losses and gains in the mechanisms that regulate play fighting could have followed either route. The best we can hope for is that the murid rodents reveal the kinds of changes, in both mechanisms and functions, that are necessary for the house mouse to make the transition from its rudimentary form of play to the complex, emancipated play of rats. More expansive, comparative work in either rodents or some other, more tractable, lineage of mammals, will be needed to determine how those changes can be implemented. But even with this limitation, we can still use the principles of the murid rodent model as a guide to see if comparable transformations can be identified in other mammals.

From Handling Pups Roughly to Parental Play

In rats, the maternal handling of pups is limited to licking them, carrying them around (from one nest site to another), standing over them so that they may suckle, and, occasionally, accidentally standing on them or kicking them. As we have seen, maternal attention, accidental or otherwise, is important for the development of the stress-response system, for its absence can lead to early brain degeneration, poor learning skills, and a host of other, unfortunate consequences. This is also true for primates.[23] In rodents, the normal interactions between the mother and her newly born offspring provide experiences for the young that help stabilize their stress-response systems. These experiences consequently lay the foundations for weanlings that are resilient and well adjusted. Then, as argued in chapter 4, these resilient and well-adjusted weanlings are in a position to make good use of the experiences of rough-and-tumble play to fine-tune their ability to deal with unexpected and unpleasant situations. Mild doses of stress appear to be important to inoculate young animals against over-reacting to events later in life. Since all mammals live in a world in which there is an expectation that infants will interact with a mother, it is not surprising that we find the same, basic experiential bedrock, with the same consequences, in both rodents and primates.[24]

However, unlike rodents, many primate mothers actively play with their offspring at an earlier stage in their development than when the infants would play with their peers.[25] Two questions arise from this difference between the two orders. First, what do these primate infants gain by having playful inter-actions with their mothers? That is, does this additional contribution by pri-mate mothers bestow further developmental advantages on their infants? Second, what is the origin of this additional, maternal contribution in primates? That is, what mechanisms are responsible for permitting female primates to engage in play with their offspring?

Our model posits that the early rough handling of infant rats helps set the range of reactions for their stress-response system, while the experiences derived from play fighting during the juvenile period provide them with a more refined calibration of how strongly they should react to their play part-ner's movements and their own loss of control in these interactions. Although maternal handling in primates lays the same foundation as it does in rats, the primate mother, by actively playing with her infant, provides her offspring with a head start in the process of fine-tuning that is only later achieved in some rodents by peer–peer interactions. Maternal play either replaces that which is afforded by peers, or ensures that juveniles will gain even more from their interactions with them. These issues will be reconsidered when we explore the origins of maternally driven play. But first, we need an empirical basis from which to determine whether mother–infant play actually improves later coping behavior.

Following on the classic experimental paradigm of Harry Harlow, who reared laboratory-born rhesus monkeys with inanimate surrogate mothers,[26] Bill Mason and his group conducted a series of extraordinary experiments that attest to the value of maternal play interactions in shaping later coping skills.[27] Infant rhesus monkeys, separated from their mothers at birth, were exposed to one of two types of surrogate cloth mothers. One type of artificial surrogate mother remained stationary, while the other mother, although identical in all other respects, was mobile. The mobile mother moved up, down, and around the cage, on an irregular schedule throughout the day. Later in life, the monkeys that had been reared by mobile surrogate mothers were more outgoing, were more likely to approach other animals and made fewer threats when they did so, were more attentive to novel social stimuli, and, when old enough, were more likely to engage in the proper sequencing of behavior necessary to copulate successfully. In all ways, unlike those infants reared by the stationary surrogate mothers, those reared by the mobile surrogate mothers behaved more like monkeys that had been reared by their natural mothers. But what does this have to do with play fighting?

The movements of the mobile surrogate mothers were not predictable; they withdrew from the infant at one moment and bumped into its rear at another. In this fashion, the mobile surrogate mothers stimulated and sustained interactions with their infants. These infants were able to withdraw from their mothers or could chase them, and then pounce on them and wrestle. Infants that had been reared with the mobile surrogate mothers initiated play fighting with their mothers three times more often than those infants that had been reared with the stationary surrogate mothers. If we think back to the organization of play fighting in rats that we detailed earlier, there are some striking parallels to be drawn between rats and the infant monkeys with the mobile surrogate mothers. As with peer play in rats, there is reciprocity – the infants and their mothers take turns in approaching and withdrawing – and, in both, there is an element of uncertainty to the interactions. These qualities make interacting with a mobile surrogate mother more exciting than with a stationary one, which probably explains why there is also a quantitative increase in the frequency of these play fighting-like interactions. Moreover, if infant monkeys are reared with a dog as their surrogate mother, this leads to juveniles and young adults that are even more socially skilled than if they had been reared by a mobile, but inanimate, surrogate mother. Therefore, there must be some subtle, additional dimensions to "mother"–infant play that go beyond those that a mobile inanimate mother can provide – even if the animate mother is a dog, not a monkey![28] Not only do these data support our contention that play fighting has an active role in shaping coping skills, but they also suggest that the early, primate-typical pattern of adding mother–infant play to the experience of infants enhances the development of coping skills that would otherwise be obtained from non-playful, mother–infant interactions.[29] One way in which this enhancement is achieved may be due to the upwelling of positive emotions that play fighting appears to foster. While hits, bumps, and any unexpected loss of control may elicit negative emotions, they are quickly overcome by the positive ones that accompany play. Research on emotional development in human infants has shown that the types of emotional experiences that are gained by preschoolers from mother–infant interactions greatly affect the child's emotional responses later in life, and can have an impact on their cognitive and social development. Mothers who express, at this time in an infant's life, a great deal of positive emotion, will influence their infants to express, later in life, positive emotions more often, whereas mothers who express more negative emotions, or neutral affect due to postpartum depression, produce children who are later more fearful.[30] What better way to enhance the mutual experience of positive emotions than through mother–infant play!

A feature of juvenile play fighting in rats supports our contention about the role of play in the development of coping behavior. In rats, play fighting involves both attackers and defenders interjecting movements into the action that decreases the performer's control over their own and their partner's behavior – this increases the moment-to-moment unpredictability of the encounters. However, this unpredictability is built on a stereotyped theme of the rats' attack and defense of the nape. That is, a theme already exists to which the rats can add small variations so as to experience unpredictability. If mother–infant play in primates already provides their young with the experience that in rodents is derived from peer–peer play only, then we should expect that primates would increase the range of variation in peer–peer interactions. Although few detailed studies on primate play are available, those that do exist suggest that juvenile primates have much more latitude in varying the theme in their play fighting than do rodents.[31]

The passage that describes the play fighting of gorillas that began this book illustrates something that a student of play in rats would find familiar. The two juvenile gorillas engage in repeated cycles of attacking and defending the shoulder region, which, if successfully contacted, is gently bitten – this is not unlike the play fighting of rats in which repeated cycles of attack and defense of the nape occur. However, there are two ways in which the play fighting of rats and those gorillas are strikingly different, and it is not the obvious one that in rats, the nape is a sexual target, and that in gorillas, the shoulder is an aggressive target. Irrespective of the origin of their play target, there are two things that gorillas do when playing that rats never do. First, when one juvenile is engaged in a play fight with another, a third gorilla may come along and join in, but may do so in a manner that seems peculiar. Instead of attacking one of the two original players on the shoulder – their species-typical play target – the third gorilla may, for example, grab onto the flailing foot of one of the other gorillas and begin gnawing on it. In the play fighting of rats, on the other hand, no matter how exposed the two rats' other body parts may be, any third rat that joins in the fray will maneuver itself in such a way so as to gain access to one or the other of the two napes, usually the one that is closer. That is, there is no generalized nosing beyond the play target in the play fighting of rats as there is generalized mouthing in gorillas. Non-target mouthing during play such as this has been reported for many primates and carnivores. In contrast, in rats and other murid rodents, such nosing or biting only migrates away from the species-typical play target by the defensive efforts of the partner, and not by a shift in the targeting by the attacker itself.[32]

Second, gorillas will occasionally insert what we term "theme-breakers" into their play fights. For example, in one incident that we observed, after repeated attacks to the shoulder, a juvenile grabbed the crotch of another juvenile with one of its hands. The recipient's startled response and subsequent leap into the air suggests that it was as surprised as we were! Where did that come from? Another example: in a typical playful attack, one gorilla will approach its partner from the front, grab its arms in its own hands, and then deliver a bite to its partner's shoulder. Sometimes, an attacker will begin following this sequence, but then may hold a dangling rope in its mouth, which will prevent it from delivering a bite to its partner's play target. It is in these ways, and many others, that gorillas seem to insert patterns of action into their play fighting that are not part of their usual theme (i.e. theme-breakers).[33]

The more complex play seen in primates compared to rats – which involves more freedom to vary the standard, species-typical, play fighting routine – may be possible because of the early occurrence of maternal play. This maternal play provides the key experiences for calibrating the appropriate degree of stress-response to unexpected events, but in a relatively safe context. Because of what is gained from maternal play, peer–peer interactions are free to be used by juveniles to expand on this calibration to a degree that is not possible in rats, and they give the young primates the leeway to explore, and so experience, more variation and subtler nuances during play. In support of this idea is the finding that great apes and humans are more versatile in their play routines than are other primates.[34] Furthermore, they also have patterns of mother–infant play that extend beyond the gentle version of species-typical play fighting seen in monkeys. The familiar peek-a-boo game that humans play with their infants also occurs in mother–infant interactions in great ape play, as do other locomotor activities that involve bouncing, throwing and swinging infants (figure 6.1).

Figure 6.1 A mother bonobo is shown lying on her back and using her feet to hold her infant overhead. From this position, the mother can jostle and shake the infant, providing it with visual, tactile and vestibular stimulation. (Adapted from a photograph by Frans Lanting. © 2008, Frans Lanting/www.lanting.com)

In our model, mother–infant play in primates provides an additional layer to the complexity of the behavior of typical, mammalian mother–infant interactions. Such mother–infant primate play may have, at first, been limited to more gentle versions of the species-typical play fighting seen in peer–peer interactions. In some lineages, however, this play may have been enhanced by incorporating forms of playful interaction that bear little to no resemblance to the species-typical pattern of play fighting as described above for great ape mothers. If we are correct, then it is apparent what infants gain from this early play, but what are the origins of this behavior in mothers?

It is likely that primates began as nocturnal, solitary living, arboreal animals. Some primates alive today have maintained this lifestyle: the lorises and pottos of Asia and Africa.[35] Unlike rodents and many other small nocturnal mammals, lorises, like other primates, give birth to one, or occasionally two, babies, at a time. A free-living loris mother "parks" her baby in some safe location when she goes off to forage, and then returns to the baby to tend to its needs. The infant, therefore, has little to no opportunity to encounter other infants. This means that, if play fighting is a necessary part of growing up, then it either has to be the mother with whom they play, or nothing. Thus, the peer-deprived youngster, if it were to have any experience of play, would have to direct playful overtures to its mother. But why should she respond? At a functional level, her playful counter-response makes sense because by doing so, she would provide her offspring with essential, developmental experiences, and this would enhance her offspring's chances of success in surviving and later reproducing. This does not, however, explain the mechanisms that propel her to play with her infant in the first place.

Adult–adult play fighting is quite common in primates and one context in which it is frequently used is in courtship behavior. The use of play fighting in courtship is particularly prevalent in primates whose social systems lead to infrequent or sporadic encounters with members of the opposite sex – that is, sex that occurs between unfamiliar animals (see chapter 5). You can see the direction in which this argument is going. The lorises and their kin have social systems that make adult–adult contact infrequent, and, yes, they have well-developed and frequent adult–adult play fighting. Indeed, the adults are so playful in some species that they play more with other adults than they do with youngsters.[36] If these mothers already have a propensity for play, then that propensity may have aided them in responding to the playful overtures of their young. In this fashion, primates representing an early lineage of the group may have deviated away from the mammalian norm of multi-infant litters and substituted mother–infant play for the missing peer–peer play. Of course, once primates adopted lifestyles where they aggregated into integrated social groups, infants would once again have had the opportunity to play with peers and not be limited to playing with their mothers. Hence, we see another multi-step, evolutionary sequence. It begins with mother–infant play substituting for the absence of peer–peer play, and then, with the addition of peers, mother–infant play expands the developmental rewards of peer–peer play.

How convincing is this scenario? If it were true, we can predict that, in other mammals that have little opportunity for peer–peer play, there should be a compensatory increase in mother–infant play. One such example is the giant panda. As pandas usually give birth to just one infant, the mother is the infant's only available playmate. As predicted from the above argument, mother–infant play fighting is very common in giant pandas. Even when pandas give birth to twins, the cubs play more with their mother than they do with each other. To complete the similarity of this animal to the loris, play fighting is also a common feature of the courtship behavior of these solitary animals.[37] Therefore, the use of play in giant pandas lends considerable support to the scenario developed for the evolution of maternal play in primates. Additional support comes from humans, where there is also a strong paternal contribution to child rearing. Father–infant play has become important in preparing children, especially boys, for the rough-and-tumble of later interactions with peers. Human fathers are even more likely to exaggerate the roughness and unpredictability inherent in play fighting than are mothers. A similar pattern is seen in gibbons, which form monogamous partnerships, where the female typically gives birth to one offspring at a time. Their offspring, therefore, seldom have the opportunity to engage in play with peers, and it is the father who primarily instigates play fighting with the youngsters.[38] As in the murid rodent model for the evolution of play fighting, then, we see new functions leading to new opportunities, which lead, in turn, to new functions. But these opportunities are only possible because of the mechanisms that are already available on which to capitalize. So far, the layer-cake model of play fighting is holding fast, but can it be extended a little more?

Returning to Some Unanswered Oddities

We can now clarify two issues by applying the logic of the layer-cake model derived from our empirical work on the play fighting of rodents. The first issue is the anomaly that we discussed in chapter 3; that overall brain size has little to say about the prevalence of play. This is surprising, because the most spectacular examples of play are to be found in large-brained species, such as humans.[39] The second issue is that only a limited number of behavioral systems have been fully developed into play. After all, most behavioral systems develop in a piecemeal fashion and so should all be equally amenable to be sources for the evolution of play. The latter issue is not one that we have dealt with as yet in this book. We will defer our elaboration of this issue until later in this chapter. Both issues are closely related to the problem of how the brain regulates and modifies play.

Why don't Big Brains Predict Playfulness?

Formal statistical comparisons within primates, rodents, marsupials, and birds, fail to show that a species' brain size can predict the prevalence of its play.[40] The evidence needs to be looked at more carefully: after all, comparing play with brain size across orders of mammals does lead to a positive correlation (see figure 3.7). It is worth considering the measures being compared more closely.

The measure for play is one that incorporates all forms of play and is an estimate of both the likely occurrence of play and its complexity. Given that good data on play are available for only a handful of species in each order of mammals, the best that this measure can capture is the maximum amount of play possible, rather than the average or the variation that may be present. As we have already seen, play fighting in rodents can range from being completely absent to present and from simple to complex. To our knowledge, no species of primate would score below rats in its play frequency and complexity. The prevalence measure used for play in figure 3.7 is a composite that captures the most exaggerated play possible in each order, but it does not tap into the range of variation present. Similarly, the average brain-size measurement to which the play measure is compared is just that – an average – again, masking the range of variation present. Furthermore, this average value is derived from only a small subset of the species from any given order – once again, potentially problematic for orders like rodents that have many species. For these reasons, within-order comparisons of values for play and brain size for individual species are more believable, and, as we have stated several times, the answer from such analyses is that brain size does not predict the prevalence of play. Despite this answer, we can still extract something worthwhile from the cross-orders comparison.

Taken at its most simple, what the positive correlation between play and brain size across mammalian orders tells us is that orders that contain bigger-brained members are also more likely to contain more playful members. This result conforms to our personal experiences – that our children and our dogs are "smarter," bigger-brained, and play more than our brain-challenged pet mice. But note that this does not explain why our pet rat plays more than our pet mouse! The core lesson to be taken from the story that we have developed here for the evolution of play fighting in murid rodents is that, as play becomes more complex, new neural control mechanisms need to be added to produce that added complexity. Strikingly, when cortical-level neural controls are added, their function is to modulate how animals use the play patterns generated by the brainstem. That is, the cortical controls do not add any new behavior patterns, but rather, modify how the ones in the brainstem do their job. As shown by rats, the really dramatic changes in the complexity of play fighting arise from having cortical systems that can "play" with brainstem-generated play behavior. The non-target mouthing and the theme-breaking maneuvers of the gorillas that were described earlier illustrate just such a "playing with play" capacity.

As in the murid model developed in this book, we can conceptualize the non-target mouthing and theme-breaking playful maneuvers as added levels of control, originating from the cortex. Mouthing is more prevalent than theme-breaking, and occurs widely among primates and carnivores (both of which are bigger-brained orders than rodents), and so may require fewer, added cortical control mechanisms than are needed to produce theme-breaking. In contrast, theme-breaking maneuvers are rare even among primates, but are quite common in the biggest-brained primates, the great apes and humans. A larger cortex, then, may afford animals greater opportunity to play with play. As another example, consider a kitten playing with a ball of yarn. It will swat it, run after it, pounce on it, and bite it. Essentially, the kitten will act towards the ball of yarn as if it were a mouse that it was attacking to kill for a meal – think of this behavior as object play that has been derived from predatory behavior. However, over developmental time, the cat will become more selective, so that the object with which it plays needs to look more and more like a mouse.[41] In contrast, consider the object play of young children. A one to one and a half-year-old child will happily pick up a phone – a real or a toy version – and pretend to have a conversation. A three-year-old child is just as likely to pick up a banana and do the same thing. That is, over developmental time, the object, for the child, can look less and less like the real thing.[42] With age, as more brain mechanisms come on-line, the two species move in different directions; one becomes more and more tied to reality, whereas the other one has more freedom to "play" with reality. The big difference between them is that the child has a substantially larger brain than the cat. A larger brain does not necessitate that play be more complex, but it allows that option to be explored if the appropriate conditions permit the complexity to be increased.[43]

Why haven't More Behavioral Systems Given Rise to Play?

This book has focused on play fighting, a commonly seen form of social play existing in many mammals and birds. Our argument has been that, by carefully examining the organization of one well-defined form of play in a closely related clade of mammals, we can gain a deep understanding of the processes that lead to the evolution of complex play, especially by taking a brain's-eye view. But the arguments that concern how the piecemeal development of behavior systems lays the foundations for the origins of play leave one big unanswered question – why don't all behavioral systems that mature in the juvenile period have play-like qualities? Why does play fighting in rats only contain behavior derived from sexual behavior, and not from agonistic behavior? This is the second issue that may become more tractable with the application of our layer-cake model. For some species, different forms of play are present and complex.[44] Even within those rodents that do engage in play fighting that is derived from either sex or aggression, the most common pattern is for one to be predominant over the other (see figure 3.5). But if we cast our net more widely across the various forms of play, rats prove to be an exception to their murid brethren, since both their play fighting (admittedly, limited to sexually-derived behavior) and locomotor play are complex.[45] A comparative analysis of whether there is covariance between these two forms of play showed something remarkable and unexpected – there is a negative correlation. In other words, as one form of play becomes more complex, the other becomes less complex (figure 6.2). This result may provide a kernel of the explanation as to why some behavioral systems do not give rise to play.

Recall that even though large brain size appears to be a relevant factor in predicting the prevalence of play when comparing *across* mammalian orders (figure 3.7), species comparisons *within* orders show either no relationship, or a weak one at best.[46] However, brain size may have an indirect effect on the patterns of play that are possible. As suggested by the negative correlation

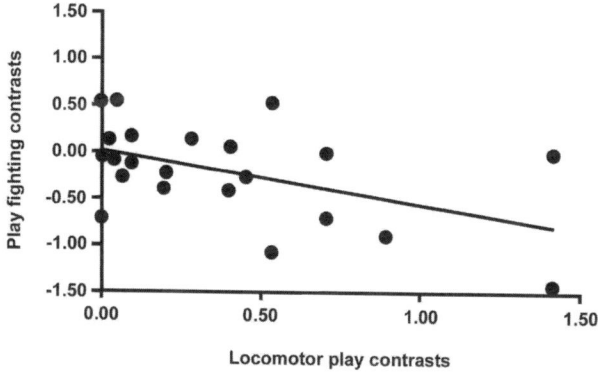

Figure 6.2 This graph shows the relationship between the complexity of play fighting and the complexity of locomotor play in murid rodents. Note the negative slope: this indicates that as one becomes more complex, the other becomes less complex. The points for this graph were converted from raw scores to eliminate the effects of relatedness among the species by using Independent Contrasts. (From Pellis, S. M. & Iwaniuk, A. N. (2004). Evolving a playful brain: A levels of control approach. *International Journal of Comparative Psychology*, **17**, 90–116. © 2004, International Society for Comparative Psychology, reprinted with permission.)

between locomotor play and play fighting in murid rodents, perhaps the brain can be viewed as finite – take up the available space building up one form of complex play and you decrease the amount of space available to do the same for another form of play. If this were so, it should also apply across mammals. Complexity in one form of play should be negatively correlated with other forms of play. Let's say the result is no correlation, rather than a negative one: this may reflect the fact that bigger-brained mammals can increase the complexity of more than one form of play before the brain space available to them is filled, and, if this were true, brain size should be correlated with an index of overall play complexity. Formal analyses are not as yet available. However, an inspection of the data compiled by Burghardt does suggest that those mammalian orders with more complex play – of any form – are also more likely to exhibit multiple forms of play, and these are also the orders with the largest brain sizes.[47]

Of course, comparison across species is needed and, as was the case for murid rodents, the ratings for complexity have to be available for all forms of play (figure 6.2). However, this approach may solve the mystery of why overall brain size does not predict play, even though it is evident that the most complex patterns of play are seen in big-brained species. To add complexity to play requires adding novel neural control mechanisms (as we have argued for murid rodents), but adding complexity to all forms of play will necessarily be limited, because of finite neural capacity. In order for different forms of play to be complex, a bigger increase in overall brain size is therefore needed so as to accommodate the required changes in neural control. Large changes in neural capacity should be associated with large changes in play complexity. The fact that great apes seem able to inject a large dose of fantasy into their play compared with most other primates (see above), suggests that this may be a fruitful avenue to explore.[48]

All Roads Lead to Rome

Before we go on to widen our exploration of play in chapter 7, there is one final issue worth considering from the comparative perspective developed from our studies of murid rodents. In rats, play fighting originally derived from sexual behavior has acquired a role in the non-sexual assessment and manipulation of social partners and opponents. There is good evidence that, for many other species, their play has also acquired this role. In many of these other species, however, the play fighting is derived from aggressive behavior. That is, even though play fighting has different origins, it has converged on the same function.[49] This leads to the question of how such convergence arose. To answer that, we need to return to the question of what makes play fighting playful. The issue of fairness was discussed at length in chapter 3, but here, we need to explore another implication of that fairness. As we discussed earlier, when comparing degus and rats, play fighting must contain a threshold level for both competition and cooperation, and, irrespective of whether play fighting is more competitive or more cooperative, every species that engages in play fighting must solve the reciprocity problem. Among species in which reciprocation arrives late in the play sequence, the tactics of attack and defense are performed in a manner comparable to how they are performed in serious fighting. In contrast, among species in which reciprocation comes in early in the play sequence, the tactics of attack and defense are only superficially similar to those used in serious fighting (figure 6.3).

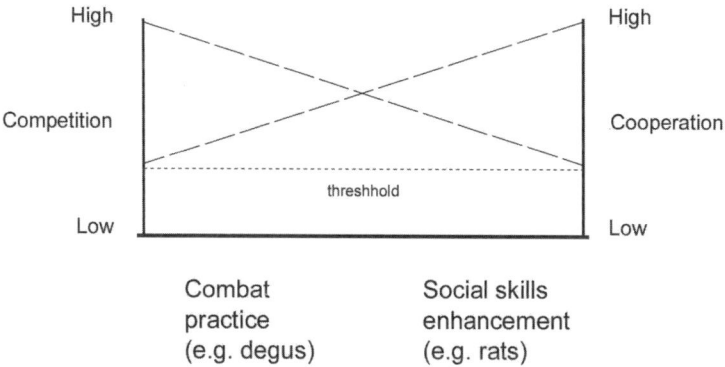

Figure 6.3 This graphical model shows that play fighting involves two components: competition and cooperation. For play fighting to remain playful, there has to be a minimum amount of cooperation. For play fighting to remain exciting, and thus rewarding, there has to be a minimum amount of competition. Within the area of the space of the graph above the minimum threshold for competition and cooperation, play fighting can range from being highly competitive (left) to highly cooperative (right). The solution to the problem of reciprocity differs, depending on which side of the model a species occupies. This also affects the kind of functional feedback that can be derived from engaging in play. Some of these differences are illustrated by noting the position in the model that is occupied by rats and degus. (From Pellis, S. M., Pellis, V. C., & Reinhart, C. J. (in press). The evolution of social play. In C. Worthman, P. Plotsky & D. Schechter (eds), *Formative Experiences: The Interaction of Caregiving, Culture, and Developmental Psychobiology.* Cambridge University Press: Cambridge, UK. © 2007, Cambridge University Press, reprinted with permission.)

The different spectrum of solutions to the reciprocity problem shown in figure 6.3 helps us to resolve a persistent problem in the play literature, and one that we discussed at length in chapter 4 – whether experience with play fighting improves combat skills or social competency skills. A species like the degu is more likely to gain combat-related experience from its play than a species like the rat. Therefore, the form of the solution arrived at for the reciprocity problem affects whether play fighting will be useful for combat training. It may be that species with play derived from serious fighting, such as the degu, are more likely to use their play fighting for combat training. But, as has already been noted, many species with play fighting derived from serious fighting nevertheless use it for social assessment and manipulation, and so the experience gained, as juveniles, from play, must have provided them with some training in social competency. The reason for this is that, even in extremely competitive forms of play fighting, subjects learn that to keep playing means to accept some pain.[50] In addition to refining their fighting skills, the subjects learn that pleasurable social interactions sometimes involve physical pain, as well as psychological pain arising from loss of control. This lesson necessarily provides the basis for making finer discriminations about social events – and so training for social competency! Whatever the origin of play fighting, it contains ambiguity – "did you mean to hurt me or not?" It is precisely this ambiguity that leads to many of the benefits gained from play fighting.[51] Because of this commonality across all forms of play fighting, many of the lessons learned from rats can be applied to people. We will explore these implications in our next chapter.

WHAT'S SO GOOD ABOUT PLAY
FIGHTING?

In the animal kingdom, there are about thirty-five phyla; that is, every species of the known million or so species of animal can be grouped into one of thirty-five distinct body plans. Armed with a catholic, but rigorous, definition of play, Burghardt has identified examples of play-like behavior in only three phyla – vertebrates, mollusks, and arthropods.[1] Within those phyla, only some lineages have been prolific in spawning playful species. One of the most prolific lineages is the vertebrate class, Mammalia. But even within this playful lineage, not all species play and, of those that do play, some do so in a simple manner and some in a complex manner.[2] Our own analysis of play fighting in murid rodents, that we fleshed out in chapter 3, illustrates this variation. From such wide-ranging, comparative analyses, we can conclude that play in general, and play fighting in particular, must have evolved multiple times. That is, when certain enabling conditions coalesce, playful behavior emerges.[3] As we have argued in chapters 4 and 6, one of the functions of play fighting in the juvenile period is to facilitate the development of social competence, and this seems to be true for both rats and monkeys – mammals from different orders (Rodentia and Primates). Similarly, in rats and in many primates, during adulthood, play fighting is used to navigate through both sexual encounters and non-sexual relationships (see chapter 5). These similarities are true even though in rats, play fighting is a modified version of sexual behavior, and, in most primates, is a modified version of serious fighting.

Thus, because they serve similar functions, behaviors that began from different origins have converged. Because of those functions, the behaviors have been modified in similar ways to serve them better.

If we compare the organization of play fighting in species from different lineages where play has converged onto similar functions, we can begin to tease apart exactly which aspects of the behavior serve those functions. Note that in chapter 3 we focused on species' differences in order to understand how lineages from a common ancestry could diverge, but now, we are turning to behavioral similarities between animals from different lineages so as to understand how the function(s) of play fighting have shaped the behavior. In playful species from many lineages, some have retained play fighting into adulthood and have co-opted it for use as a means of social assessment and manipulation.[4] But what is it about play fighting that makes it useful for such a purpose?

Quite simply, play fighting contains an element of ambiguity – there is always the risk that a punch, slap, kick, or throw was deliberately intended to be more painful than the receiver of that action had expected. Because it is play, one's partner has to be given the benefit of the doubt, but how many transgressions have to be tolerated before it is concluded that the partner is taking unfair advantage? We believe that this gray zone of uncertainty endows play fighting with its value as a tool for social assessment and manipulation. Depending on the nature of the relationship, taking advantage of the situation in a playful context can tell one much about the robustness of the relationship; alternatively, it can be informative about the weakness of one's social partner, and, hence, how ready the relationship is for a change. Again, as they have done so many times in this book, rats provide a good model that we can use to understand this situation. We can therefore re-explore the uses of ambiguity in the play fighting of adult rats, and then, from that vantage point, we can expand the view to include other mammals, including humans.

The Uses of Ambiguity

Adult male rats use play fighting-like behavior to facilitate copulation during sexual encounters, as well as to assess and manipulate social partners in nonsexual, dominance-related contexts. Within the colony, subordinate males use play fighting as a means to maintain friendly relations with the dominant male and, during both within- and between-colony encounters, males can use play fighting to assess the dominance-holding capabilities of other rats. It is the play fighting in these non-sexual contexts that most closely resembles the play fighting seen in juveniles. Initially, this seems surprising, given that play fighting in this species involves elements of sexual behavior, but as we described in detail in chapter 6, precocial sexual behavior in the ancestral lineage was likely modified and co-opted into non-sexual contexts. Many other species of mammals besides rats have also been reported to engage in adult–adult play in similar contexts. However, unlike rats, some species have exaggerated the use of play fighting in courtship, and, in some lineages, play fighting has become either integral to courtship or is an option available under certain conditions. An example of each may better illustrate these two conditions.

Pottos are nocturnal, tree-living primates. Like lorises, which we discussed in chapter 5, male and female pottos maintain independent home ranges, forage independently, and generally sleep by themselves. This pattern of behavior is thought by many researchers to resemble closely the ancestral condition in the primate order.[5] While antagonistic to territorial intrusion by members of the same sex, males will visit females that live in territories that overlap their own. A male will intrude into the territory of a female, night after night, until she becomes accustomed to his presence. At this juncture, she will allow him to groom her. Over successive nights, the grooming not only becomes prolonged, but also, reciprocal. With yet more nights and more mutual grooming, the pair begins to engage in play fighting. They hang upside down on a branch by their feet and grapple with their forelimbs, grabbing and pulling at one another (figure 7.1). After several nights of this grooming and grappling play, the pair may copulate. The sequence of ever more intimate handling of one another seems to be necessary to overcome the antagonism that individuals exhibit toward intruders into their territories. Play fighting, then, is a crucial component of potto courtship.[6]

Moving onto a somewhat larger mammal, a male grizzly bear, during the mating season, may restrict the movements of a female to a secluded location, away from other bears. Over several days, they will make periodic contact, with some of that contact involving head-to-head wrestling and pawing – that is, play fighting. Whether such contact occurs, and for how long it proceeds, seems to be regulated by the female. The play fighting of the bears appears to be an important means by which to reduce the aggression between these otherwise solitary animals. Although this seems similar to the case of the pottos,

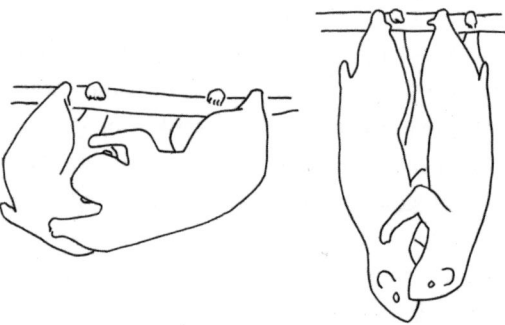

Figure 7.1 Grappling play in a male and female potto. They grab, hold, and compete to groom and nuzzle one another. (From Epps, J. (1974). Social interactions of *Perodicticus potto* kept in captivity in Kampala, Uganda. In R. D. Martin, G. A. Doyle, & A. C. Walker (eds), *Prosimian Biology* (pp. 233–244). University of Pittsburgh Press: Pittsburgh, PA (Fig. 6, p. 239). After a good faith attempt to do so, copyright could not be determined.)

there is one critical difference. When male–male competition is more intense among grizzly bears, especially in high-density populations, this prolonged building of a temporary sexual bond does not take place; instead, males engage in more coercive patterns of courtship. Thus, for some species, using play fighting as an element of courtship is a conditional strategy, one that is used only under certain conditions.[7]

Play fighting is also reported in non-sexual contexts for the adults of many mammals, and these fall under two general categories: to build and maintain strong social bonds and to gain advantage over a social partner by changing the nature of the relationship. Like adult rats, many species of mammals, including greater bush babies, common marmosets, slow loris, and New World deer, use play fighting to establish and maintain social bonds and, again, like rats, many species, including chimpanzees and Old World deer, use play fighting to test existing or emergent relationships.[8] Whether these affiliating and testing functions of play fighting extend to interactions with novel animals, such as an intruder into the colony, seems to vary across species. As we have already noted, adult rats use play fighting both for affiliation with the dominant colony member and to test for opportunities to reverse dominance relationships within the colony, as well as to determine whether dominance can be established when unfamiliar adult males are encountered. It is intriguing that both the slow loris and the greater bush baby use play fighting within their colonies for social affiliation, but that only bush babies use play fighting in a rat-like manner when encountering unfamiliar animals in a neutral arena. Similarly, even within-colony use of play fighting can vary across closely related species. While sub-adult spotted hyenas appear to use play fighting for affiliation, and hence, for integration into the clan (i.e. to establish and maintain social bonds), sub-adult brown hyenas preferentially engage the adults in play fights, indicating that they are probing for a position in the clan's dominance hierarchy.[9]

And on to Humans

As occurs in rats, humans can also use a playful context to establish and test social relationships, by providing a relatively safe setting within which to explore them. If one oversteps the boundaries tolerated by one's partner, then "sorry, I was only playing" provides a graceful exit. Not unexpectedly then, in humans, the problem of overcoming initial reticence or unfamiliarity in the early stages of courtship can be solved by playfully flirting; this can involve wrestling, gentle hitting, and tickling. In such courtship play, women seem to tolerate a narrower range of acceptable playful contact than do men, even though with respect to their own behavior, they allow themselves a broader range of intensities of contact.[10] This latter observation highlights something important about using play fighting in a socially manipulative manner. The boundaries of acceptable play behavior can vary across partners and contexts. Thus, the use of play fighting as a means for assessing and manipulating partners can be very taxing both emotionally and cognitively, as actions need to be evaluated continually. Two important questions arise from this: first, are there alternative ways that playful interactions can be used for such social testing? Second, are there experiences early in life that can lead to the improved use of playful ambiguity? We will deal with the second question later in this chapter, and for now, examine only the issue of the forms that play fighting can take.

As well as the ambiguity inherent in play fighting, another of its critical features is that it involves physical contact. We have already discussed how making gentle physical contact can be quite calming because it induces the release of endogenous opioids, but it can also be informative. A partner's response of "no" to the question "are you still angry?" can be assessed for its honesty if one's hand is resting on his or her shoulder when the verbal response is given – tense muscles are an obvious indication that the relationship needs more work.[11] So, the physical contact inherent in play fighting is both rewarding and informative, but is it essential? For humans, perhaps, playful ambiguity can be divorced from the need for touch. The sizes of the social groups in which humans live are staggeringly large from the point of view of other primates. The average primate devotes some period of time each day to physical contact with other members of its social group, using a combination of grooming, huddling, and playing. This physical contact can provide a means by which to keep track of social relationships. But, as Robin Dunbar notes, the ability of humans to monitor and maintain relationships by physical contact only is simply impractical because of the large size of their networks. Instead of physical contact, humans have developed another trick that can work to cement relationships – we talk! Indeed, over 90% of the content of the conversations between people is gossip, even those of the most highly educated humans. So much for "highbrow" discussions about critical world events – people are mostly talking about who is doing what to whom. Usually, the objects of our gossip are people we know personally, but in our modern, media-crazed world, we are also happy to gossip about those whom we only know vicariously, such as movie stars and sports figures. This gossip serves two purposes: it can induce the warm feelings that come with sharing "secret" information with another, and it can also provide potentially useful information about possible competitors.[12] In some ways, talking has the same function for humans that physical contact has for many other social animals: a sharing of both intimacy and information. One form of language use, humor, also resembles play fighting in its most critical feature – ambiguity.

Humor is closely related to laughter. Laughter in humans may, in turn, be comparable to the open-mouth facial expression that is typically present not only in the play fighting of other primates, but also in many other mammals. From a functional point of view, laughter has been shown to activate the pleasure centers of the brain and induce a positive state in those laughing, and sometimes, even in those that witness others laughing. Not surprisingly, laughter is often used when confronting uncertain social situations, probably as a vehicle with which to induce a positive mood and avoid misunderstandings.[13] However, humor – the major instigator of laughter – is also employed to reduce stress and facilitate bonding, as well as to gain status. Thus, using the "I was only joking" gambit is not unlike the "I was only playing" ploy. The give-and-take of jokes, which are often delivered at the expense of someone else, is an effective way with which to probe for the strength of relationships and for the weaknesses that can be exploited. Again, the similarity to play fighting is remarkable. A further similarity between the two is that, just like play fighting, humor is more often used by males, both as boys and as men, even though there seem to be few differences between the sexes in their ability to appreciate humor.[14] It is conceivable, then, that with the advent of spoken language, humans developed a verbal, non-physical form of play fighting, a form of social interaction that is intimate, informative, and ambiguous. The use of a formalized, but playful, verbal banter has been documented in many African-American communities as a means by which to reinforce valued social relationships and probe suspect ones. It seems highly plausible that, depending on the cultural milieu, the functions of speech that either enable social bonding to occur, or which confer a social advantage, have been promoted. An excellent example of the use of verbal banter, but with a sharp edge, for elevating one's social status – social climbing – is shown in the 1996 film by Patrice Leconte, *Ridicule*, in which courtiers, so as to gain favor in the court of Louis XVI of France (1780), poke fun at other, more dominant, courtiers, but do so in a manner that allows them to gain the better of one another without it being too obviously vicious. The film most ably demonstrates that the boundary between playful joking and insult can be very narrow and that only the most skilled are capable of success.[15]

Harking back to our layer-cake model for the evolution of play fighting that we developed in chapter 6, we can identify some of the potential building blocks that led to human humor and laughter. First, there are the facial gestures that are associated with play fighting. Second, these gestures eventually evolve the capacity to arouse, independently, the same feelings as play itself. Third, verbal banter takes over, or adds to, the modes of relationship building and testing that are based on touch. Fourth, verbal communication becomes sufficiently sophisticated so as to contain all the elements that are essential for contact-based play fighting: rewarding behavior that is informative but suitably ambiguous to be used to navigate and test social relationships. The first two steps are consistent with the evidence that we have already explored, which indicate that the neural mechanisms that make play possible and pleasurable are both present and similar across all playful mammals – at least, for the small handful that have been studied. The next steps are purely speculative, since there are no extant species that fill the cladogram between chimpanzees and humans. Nonetheless, this speculation can be useful in helping direct researchers to identify where in the brain to look for those uniquely human attributes.

We would also be remiss if we did not point out that there are hints of humor in our closest living relatives, the chimpanzees. In one example, a juvenile chimpanzee that was living in a colony maintained on a ranch in the USA was seen to hide under a parka, squat down, and crawl towards a herd of horses in a corral. The horses did not appear to take any notice of this movement. Once at the edge of the corral, the chimpanzee jumped up, removed the parka, and flayed it about, and the horses scattered in panic. The chimpanzee ran off, while simultaneously making panting noises, similar to that of a human laugh, and with an associated play face. Once the horses had settled down, the chimpanzee then repeated its performance.[16] We certainly know children who would think these antics great fun!

All joking aside, even with laughter and humor as uniquely human derivations of mammalian-typical play fighting, we also know that contact-based play fighting is used by adult humans. A version of play fighting that is more recognizably like that of pre-adolescent children has been well studied in post-pubescent children, especially in boys, where it is employed both to assert and gain dominance.[17] As shown in our studies of rats, dominance can be negotiated through play fighting; this means that these animals do not have to resort to full-blown fighting to settle any problems in their relationships. Whether it is employed by rats, hyenas, monkeys, or humans, play fighting offers a level of nuance in social interactions that permits differences to be sorted out without the use of extreme violence. It would seem that adult animals that are better skilled at using the ambiguity inherent in play fighting will also be better at negotiating intricate, social worlds. But before we discuss how these skills are developed, we should reiterate that, under certain situations, humans, as well as some non-human animals, are able to use play fighting to gain social advantage. Of course, this means that there are losers as well as winners, whether the game is played "nicely" through play, or "nastily," through more explicitly coercive means.

This darker side of playful fighting was demonstrated well by Brian

Sutton-Smith, in an historical survey of play in the school grounds of New Zealand.[18] He found that, on occasion, a playful overture offered by one child could be used as a pretext to torment another. There is little doubt that this experience will resonate with many a reader. Indeed, one of the authors of this book (Sergio) spent the first two years of his secondary schooling (7th and 8th grades) in Australia at an all boys' school, which is not always the best environment in which to be a loner. One group of young toughs took great delight in initiating the new boys to the ways of the school playground. Under the guise of a smiling face and a playful demeanor, one of the bigger boys would throw you to the ground – usually in a grassed area – and then pin you on your back so that you could not move. He would then proceed with the "rats" treatment, which involved repeatedly tapping you, with his forefinger, and with an uncomfortable degree of force, on your chest, while, at the same time, shouting in your face, "rats, rats, rats…" in quick succession. Any complaints or threats of retaliation by the victim simply led to one being given a more prolonged treatment of "rats." There was no way out of this fix but to smile and laugh while simultaneously enduring it – that is, to pretend that you accepted this treatment as an act of play. The longer-term solution to such "playful" bullying involved two, complementary strategies. First, you picked fights with the smaller members of this marauding gang of young toughs, especially at the times in which they were isolated from their gang. This established your reputation as to the limits to which you could be pushed. Second, you formed a gang of your own, with a sufficient membership of boys who were tall and bulky – and hence, intimidating – for their age. This latter strategy inhibited the other gang from picking on any one member of your gang, especially if you always made sure that you traveled in the school playground with three or four of your own gang members.[19]

Another important issue discussed by Sutton-Smith is well illustrated by this tale of "rats:" that overt facial expressions and other behavioral signs of playfulness are not necessarily foolproof ways of distinguishing playful from non-playful fighting. We can assure readers that Sergio did not enjoy being repeatedly tapped on the chest, no matter how much he smiled. Fortunately, in the case of children or young adults, observers can ask the participants whether they had been engaged in a serious or playful fight.[20] Even though people are not always truthful about their behavior, asking the participants may provide additional information that would enable researchers to make a more sound judgment. But when it comes to non-human animals, we are denied this additional information and so are forced to use only the overt, observable behavior. So, how do we know for sure that any particular play fight is truly regarded as such by both participants? We do not. To make such a judgment, we need detailed knowledge about the particular species' behavior and knowledge about the individuals involved and their relationship.[21] Based on intimate knowledge such as this, Clara Jones was able to report that, among adult howler monkeys, higher status individuals used play fighting as a means by which to keep monkeys of a lower status in their place.[22] Thus, a non-damaging form of behavior is used to remind other members of the social group who is the boss, and is done without having to resort to a serious fight that could force a subordinate to retaliate. The situation is much like that which we discussed earlier concerning jokes: the more dominant member of an office can make the occasional joke at a subordinate's expense, and in so doing, is reminding everyone, albeit in a playful manner, exactly who is in charge.

That there is a dark side to play should not come as a surprise – after all, play fighting is being used in these instances as a means of social assessment and manipulation. Therefore, we should expect that play fighting will span the spectrum from positively reinforcing valued relationships to coercively manipulating social rivals. Play fighting thus offers its users a wide range of options in trying to adjust social relationships to their advantage. The choices that we have are not between everybody loving one another and living together in egalitarian bliss versus living in dictatorial brutality, but rather, between convincing someone to give you a share of the pie or taking it from them whether they want to give it to you or not. In democratic, law-bound societies, overt brutality in gaining what we believe to be our just desserts is unacceptable and is likely to elicit fierce resistance. Instead, finding ways of getting others to cooperate with you is the socially and legally acceptable route to follow. In this context, the wider the range of alternative strategies available to an individual, the better that individual will be able to navigate the complex social networks of large, democratic societies with their global markets. If the story that has unfolded for rats is any guide, our suspicion is that experiencing play fighting as juveniles is crucial for the development of the nuanced social skills so necessary for success in the modern world.

How to Become Good at being Ambiguous

When asked to discriminate between instances of playful fighting and serious fighting, children are more accurate than adults in determining this, and, in adults, men are more accurate than women. However, if women have had personal experience with play fighting as children, they are better able to discriminate than those who lack such experience. Clearly, experience can make a difference.[23] To our knowledge, there are no studies available that have examined whether differential experience of play fighting in humans leads to more subtle discriminations of playful encounters in adulthood. However, there are some studies that indicate that children who have had more experience with play fighting are also better able to solve social problems. Further, the experimental data on rats that we have extensively reviewed in the previous chapters suggest a direct, causal connection between play fighting and social competence.[24] Given that we have also reviewed experimental data on non-human primates that paint a picture similar to that for rats, it is not unreasonable to suspect that the same should apply to humans. However, there is a persistent problem with how best to interpret childhood play fighting that we believe unnecessarily confounds the issue of whether experience with play fighting promotes social competence.

There seems to be widespread consensus that play fighting in post-pubescent children is organized and used so as to compete for dominance, and, at least, in male–male dyads, play fighting in adolescence does appear to have a rougher quality, making it more like a form of quasi-aggression. However, when it comes to pre-pubescent play fighting, there is less agreement as to whether it functions as a means of combat training or improves social competence.[25] Reconsideration of how play fighting unfolds will help resolve this conflict. As we discussed in chapters 3 and 6, play fighting must contain a threshold level of both competition and cooperation, and irrespective of whether play fighting is more competitive or more cooperative, every species that engages in play fighting must solve the reciprocity problem in order for it to remain playful. That is, players need to exercise fairness in their play fighting. In species in which reciprocation comes late in the play sequence, the tactics of attack and defense are performed in a manner comparable to how they are performed in serious fighting; in such species, play fighting can serve as practice for adult aggression. In contrast, in species in which reciprocation comes early in the play sequence, the tactics of attack and defense only super-ficially resemble those used in serious fighting. In such species, play fighting is a poor means of practicing serious fighting. Examples of the former include degus and pigs, and the latter, rats and many primates (see figure 6.3). Although there are some good, descriptive analyses of play fighting in human children, they are of insufficient depth to determine clearly whether their play fighting is more like that of the degu or the rat.[26]

That personal experience with play fighting can influence one's ability to discriminate between the playful and aggressive behavior of others suggests that, even if a child's experience of play fighting does enhance later fighting ability, it must also change their ability to differentiate between social actions and weigh them appropriately. Thus, even an extremely competitive form of play fighting appears to have an in-built mechanism for such social training – subjects learn that to keep things playful not only requires that they accept some pain, but also accept that their partner must have the opportunity to win.[27] That is, in addition to refining their fighting skills, subjects learn that pleasurable social interactions sometimes involve physical pain, as well as the psychological pain that arises from loss of control. Therefore, no matter whether play fighting in the juvenile period is primarily designed to promote the development of social skills or combat skills, the subject's ability to make more nuanced discriminations about the social action of others should be enhanced. Therefore, whatever else play fighting is good for, it is clear that it promotes the development of social skills. This, in turn, has implications for how we, as a society, deal with play fighting in childhood.

Current textbooks on children's play devote very little space to the discussion of play fighting – often only in the context of whether it is really play. Even a recent book aimed expressly at promoting play in childhood does not contain the terms "play fighting" or "rough-and-tumble play" in its index.[28] There are several reasons why this neglect is unfortunate. First, play fighting has been shown to represent between 5–10% of the spontaneous play occurring in the school playground. Second, despite variations in the imagery component of play fighting across cultures and historical times, play fighting is remarkably similar across the world. Third, it is the one form of human play that most closely resembles the play of non-human animals.[29] That it occurs spontaneously, is consistent across cultures, and is comparable to the play of other species, means that this is a form of play in humans that is likely the most tractable as an object of study, and the one most likely to offer insights into forms of play that are unique to humans, and are, contextually and culturally, more variable. Not the least important point is that the experiences gained through play fighting as a juvenile are very likely to have consequences for the skills that, later in life, are related to social competence.

Toward the end of the 1990s, when our work on play fighting and that of many others around the world was leading to the inexorable conclusion that, in rats, juvenile experience with this behavior was crucial for the development of social competence, we had an epiphany. It occurred on a visit to Paris, France. After having spent at least half of our adult lives in North American cities, we were accustomed to seeing children spending their leisure time mainly in shopping malls, and, when outside, under the supervision of parents or teachers, or engaged in organized games, such as football. In Paris, as in many European cities, most people live in the central part of the city in small apartments, as compared to the houses of North America. It is said that Paris is but a collection of villages, and in the modern city, many of these "villages" come with their own verdant little square. One day, we were out walking at around 4 o'clock in the afternoon, when we passed one such place, the Place des Vosges, and noticed that parents and guardians were bringing their children, of primary school age, into the park. We entered and sat down on a bench, and saw, very quickly, what was happening there – the adults sat around the periphery, chatting with one another, and the children congregated in groups on the grassed area. Some children were playing ball games, others had model cars or planes, and some had skipping ropes. Still others ran wild, chasing each other, tackling one another and, yes, play fighting. Occasionally, a child would appear to land hard on the ground, and then would stand, and begin to cry, while at the same time, turning to look in the direction of whom we assumed to be a parent or guardian. The child would quickly recover and again join the fray. We had almost forgotten what it was like to see children playing like children and engaging spontaneously in myriad forms of playful endeavor. Their caretakers were there, providing security if a child were hurt or was unfairly treated by another child, but, otherwise, it seemed that the choice of games and play partners was decided solely by the children themselves.

Our observation of children being allowed to engage in spontaneous play fighting with varying degrees of intensity is at odds with the ever growing number of newspaper stories with which we had become familiar in North

America, where more and more schools and public parks are banning any form of roughhousing by children, whether play fighting or chasing. Fear of injury and the threat of litigation are no doubt reasonable concerns of those responsible for the well-being of children and so there has been an escalation in the banning of any form of unstructured play.[30] Many school districts have already eliminated recess, even though the little empirical evidence that does exist indicates that giving children periodic opportunities for unstructured activity throughout the school day enhances their scholastic performance, rather than detracts from it.[31] Given the ever-growing rates of obesity among children, such restrictions are likely to be deleterious in many ways, but our main concern here is with the potentially negative impact that they would have – or are having – on the development of social competence. Of course, this is not to belittle legitimate concerns about injury, bullying, and litigation. Nonetheless, the French example described above offers a compromise – adults are present offering a safety net if one is needed, the play is concentrated on thick mats of grass, so minimizing the likelihood of serious injury, and there are sufficient children available for individuals to select, for themselves, activities and partners that are best suited to their predilections. These features for making play opportunities safer are the very ones that are starting to be recommended by educators who are concerned about the ever-diminishing opportunities for children to engage in play fighting.[32]

Other options could include the development of games, such as computer-based ones, that engage the skills that are essential for children to learn in play fighting, such as turn taking, and thus fairness. These games could possibly train children in these crucial skills, but, at the same time, would avoid their risk of injury.[33] While this option has its merits, it does not provide the vigorous exercise effects that are associated with play fighting, nor does it include physical contact, which, as we saw earlier, may enable critical, social bonding experiences and information gathering to be had. These examples illustrate that there are ways of minimizing the risks while not eliminating the important experiences that arise from play fighting. It is imperative that we understand how the experience of play fighting improves a person's later abilities both to navigate through ambiguous social interactions and to use ambiguous patterns of interaction to good effect. Even though we do not know as yet the exact neural mechanisms that are involved in play fighting, we do know from observational studies on children and experimental studies on non-human animals, that it would be wise if we, as a society, found ways to allow our children the opportunity to engage in this behavior. Before leaving our discussion of the role of play fighting experience in the juvenile period of development, we will just touch on one more way in which our knowledge of rats may be helpful in resolving otherwise perplexing issues.

Why is Play Fighting More Frequent in Boys than Girls?

A major problem with the hypothesis of play fighting as a "tool for refining social competence" is that women are more socially competent than men, yet they engage in less play fighting as children.[34] However, this theory may not only be salvaged, but may even be reinforced by this sex difference. Several facts need to be considered. First, play is a context-sensitive behavior. It is most likely to occur in resource-rich environments and is less likely to occur in resource-poor environments.[35] So, play has to be a contingently adaptive strategy – that is, it cannot be critical for development, but rather, when it is available, it is helpful. Second, males are generally more aggressive, and often, the most dominant males are the ones who are most aggressive, and so the least sensitive to social nuance. Third, females are, generally, less physically imposing than males and so rely on more subtle social skills than males to meet their needs.[36] Based on these sex differences, it would seem reasonable to predict that in females, those brain circuits that are needed for social competence are already developed, independently of social experience, to a greater degree than in males. Recall that in chapter 4 we showed that play fighting experience in rats in the juvenile period led to alterations in the prefrontal cortex of the brain, with neurons of the OFC becoming more complex and those of the mPFC less complex (figure 4.7). Also recall that we cited evidence from neuroimaging studies on humans that showed that whereas the OFC is involved with cooperation, the mPFC is involved with competition (see chapter 4, endnote 82). In female rats, the OFC is more complex and the mPFC less complex than it is in males, whereas in male rats, the opposite is true, with the mPFC neurons being more complex and OFC neurons less complex. These sex differences are dependent on exposure to gonadal hormones during early development.[37] The core, sex-typical biases are thus built into each sex in ways that are not likely to be contingent on social experience.

Of course, we have already examined evidence that play fighting experience in the juvenile period can alter social competence and so, presumably, the brain mechanisms that mediate this, but what these findings also suggest is that processes are in place to ensure that females have a brain that is more capable of coping with the social world, and males have one that is more capable of competing with rivals. One reason for such a bias is that it is likely, generally, that a female's reproductive success is dependent on her social competence, be it in her interactions with other females, with males, or, most certainly, with her offspring. In contrast, for males, reproductive success is more variable, with social skill being only one of a possible number of assets that are critical. There is evidence from studies of an increasing number of species that indicate males may adopt any one of a multitude of strategies so as to gain sexual access to females.[38] For example, while large, aggressive, and dominant males may have priority of access to females, subordinate males may use more subtle techniques to gain some reproductive opportunities. This is well illustrated by the male "seduction" of females in orangutans.

The largest, most dominant, male orangutans in an area will climb to a high point on a tree in their territory and emit what are termed "long calls." The sound of these calls travels vast distances in the forest, notifying receptive females of their presence. Female orangutans find these calls appealing and will travel to the caller's location for an assignation. Because males that are small in stature cannot fake long calls, they have to resort to using different strategies to gain access to females. The smallest adult males in the area may follow a female, and, for prolonged periods of time, interact with her playfully. This may lead to her allowing copulation to take place. Moderately sized males may resort to intercepting females as they travel to the attractive dominant caller and use their superior body size and strength to force copulation.[39] Depending on the male's size and status, he may demonstrate little social finesse, brutal social behavior, or friendly persuasion. Except for certain extreme situations, it is not possible to predict the kind of social context in which a male may find himself. In one social context, an aggressive approach may work, but in another, a more subtle, socially sophisticated one may be needed. Aside from size and age differences, animals also vary in their temperament, which, in turn, affects which of these strategies they adopt.

The most thoroughly studied temperamental feature is that of the "boldness–timidity" gradient, a dimension along which individuals from a wide range of species have been shown to vary. Furthermore, such temperament differences have been shown to be consistent across situations, so that if an individual is bold in one context it will also be bold in another. Finally, some of the underlying genetics and neural mechanisms for such variation in temperament have been characterized.[40] For example, among male rats, bolder, more aggressive males are more likely to become dominant than timid ones and are also more likely to exhibit threat postures. Timid rats, on the other hand, are more likely to avoid other males and freeze in their presence. Further, as juveniles, bolder rats are more likely to initiate play.[41] Of course, if a highly timid male rat is housed with a very bold male rat, it is no surprise that the former will quickly adopt the role of subordinate and the latter the role of dominant, but what if a bold rat is housed with a bolder one, or a timid rat with a more timid partner? In the real world, where most of those encountered are likely to be only a little different from oneself, any individual may have to be able to accommodate its social behavior to the context, and this is what we have seen countless times when male rats are placed in a situation where they have to form a relationship with other male rats.[42]

Therefore, some degree of social competency is a requirement for most animals. However, it should be borne in mind that it is the weaker or subordinate animal that has to be more attentive to the actions of the stronger or more dominant one. If sufficiently powerful, the latter can be oblivious to those that surround him or her. In the last few years, because of the repeated television footage, the blunt power exuded by the former Iraqi dictator, Saddam Hussein, is an image that readily comes to mind. He was often shown walking through a crowd, nonchalant, while all those in the crowd are fawning over him. The trouble is that it cannot be predicted, early in life, whether an individual will become a Saddam Hussein or one of his weaker vassals. It is likely that in his climb to the top, Saddam did a lot of fawning of his own.

In conclusion, a male's need for social subtlety may vary depending on his status. Under poor resource conditions, the premium for success is to grow as quickly as possible, and doing so may require being as mean as possible so as to out-compete rival males. Under these conditions, it is attitude and strength (or size) that counts, and so prefrontal mechanisms that are critical for regulating competitive ability are more useful than a prefrontal cortex that has the capacity for a lot of cooperation, and thus, social nuance. Besides, in such a situation, play is a luxury that can be ill afforded, and so it may not be available to alter the balance of mechanisms in the prefrontal cortex. In contrast, in a resource-rich environment, most males will likely end up having a high-quality phenotype – where all the males are big and show a similar level of competitive ability. In this situation, improving one's non-aggressive competitive skills becomes more important, and to be good at social manipulation in the absence of using overt aggression requires bolstering the capabilities of those parts of the prefrontal cortex that regulate a cooperative form of social competence. During the juvenile period, plentiful resources trigger an increase in play, which, in turn, redirects development to increase investment in the social brain, which then leads to improved competitiveness via non-physical means. Essentially, the model we propose suggests that females are biased to develop a socially competent brain, whereas males are biased to develop a brain that is minimally socially competent but if the conditions are right, can use play in the juvenile period to develop the social brain further. This model explains three facts: (1) that males play more than females; (2) that females are more socially competent than males; and (3) that the degree of play in which juveniles engage depends on local conditions.

There is variation in the degree to which play fighting by children is tolerated in different cultures around the world. In some, play fighting by children is not merely tolerated but even encouraged; in others, it is positively discouraged.[43] Societies also vary in the rigidity of male hierarchies and therefore in the kinds of skills that enable individuals to gain status. We hazard a guess that, in societies in which males succeed in furthering their status only via physical prowess or non-social activities, social play will be dampened or, if allowed, will emphasize competition. In contrast, in more egalitarian societies, in which convincing others to cooperate or where being able to detect subtle social maneuvers by competitors is important for social success, then not only should play fighting be tolerated, but its cooperative component should also be emphasized.[44]

Although there may be important sex differences in the requirements for social competency, it is very likely that both sexes require at least a minimum amount of play fighting experience in the juvenile period so as to fine-tune or calibrate their assessments and reactions to others' social actions. Richard Tremblay and his group have shown that, contrary to what had been thought for a long time, people are at their most intensely aggressive as young children, peaking at around two years of age. Then, in the course of the next two or three years, that propensity for aggression is brought under control.[45] Indeed, children that fail to restrain their aggressiveness by this age tend to continue to be overly aggressive for life.[46] Reciprocal playful exchanges with one's mother, followed by boisterous play fighting experiences with one's father, seem essential to curtail early aggressiveness. Failure to engage in such play appears to lead to an inability by the child to regulate reciprocal exchanges with peers properly, making it more likely that such children are ostracized from playing with others, which then leads to more frustration and more aggression. Thus, by these means, play fighting that begins with one's parents and continues with one's peers, provides a means by which a child can learn to regulate their aggressiveness.[47] You may recall, for example, that when your pet kitten or puppy first began to play with you, it had little restraint in using its claws or even biting.[48] It was only by chastising it and, perhaps, by smacking it lightly on the head or issuing a downright refusal to play that it learned to hold back the use of its claws and teeth. Again, we are drawn to the conclusion that play fighting has an element of "make believe," and that the participants must accept that what transpires during such play is not to be viewed as aggression, but that, because of its inherent ambiguity, some transgression is acceptable. The problem arises when a young animal, be it a human child, a monkey, or a rat, consistently fails to restrain its actions, and so finds that it is abandoned by its play mates.[49] As there may well be no substitute for play fighting as a means by which to learn these critical lessons of fairness, for both sexes, a certain minimum amount of play fighting may be necessary.

To sum up, what we have shown is that there are general properties of play fighting, such as its inherent ambiguity, that have been co-opted by many species to evolve a form of social interaction that can be used to probe social relationships. The advantage of such playful probing is that it can avoid escalation into vicious conflict because a fail-safe exists – the "sorry, I was only playing." We also saw that to become skilled at such playful probing requires practice, especially as juveniles. The unpredictability that exists in play, particularly when coupled with its inherent ambiguity, makes it a fertile ground for training individuals to cope with the unexpected. Of direct relevance to the theme of this chapter is Tony Pellegrini's observation that children that experience the most varied social play are also the most adept at developing novel solutions to social problems.[50]

With regard to the use of playful ambiguity in adulthood, an unresolved question remains: do all species that use it have comparable neural mechanisms? We suspect that, for mammals, at least, this is the case. That is, cortical mechanisms that are able to modulate the content of play fighting in rats are also likely to be the same as those that do so in humans. Comparative studies with rats, monkeys, and humans should resolve this matter.[51] With regard to how the skillful use of playful ambiguity develops, the major unresolved question concerns what it is exactly that needs to be practiced. As we discussed at length in chapter 4, there are many routes, some direct and some indirect, by which experiences early in development can affect skills later in life. It is this issue, of what is practiced by play in the juvenile period, and how it is done, that will be the focus of the final chapter. After all, it is appropriate that we end this book by providing a vision of where the study of play may lead. As history buffs, we also know that the past is often a good guide to the future – thus, by understanding what the arguments of the past were, we are often able to sharpen our appreciation of the data we have in the present and so formulate the questions that need to be tackled in the future.

8

FROM HERE TO ETERNITY

We will end this book by tackling what we believe to be the most challenging task that lays ahead for the study of play. As this is the final chapter, we also want to make sure that the reader understands the particular approach we have taken in our own journey in studying such a complex phenomenon. As illustrated by the very structure of this book, we believe that when confronting such a phenomenon, it is necessary to engage in a two-step process: the phenomenon needs first to be fractionated into its components and then reassembled. Taking short cuts may make the task more palatable, but may also lead one to dead ends. If the reader has managed to work their way through the detail-intensive chapters 2 and 3, they would then be well placed to understand our logic in the subsequent chapters, and, hopefully, in a position to identify any flaws in our arguments. So, before our final exploration of play, let us make explicit some of the analytical tools that we think any student of complex behavior would be well advised to consider.

The Deconstruction of Complexity

As we have seen, several analytic steps are necessary to try to explain how a rat-like species with complex patterns of play fighting could have evolved from a mouse-like species with little or rudimentary play. First, we identified a practical model species. Second, we carefully described the play fighting of that

model animal. Third, using cross-strain comparisons, between-sex comparisons, across-age comparisons, and individual variation, we characterized the
organization of play fighting. Fourth, we then placed into comparative perspective the components of play fighting and its organization in the model
species by examining how they differed within a cluster of closely related
species. Fifth, we identified the species-level changes necessary to transform a
non-playful animal into one with a complex pattern of play and mapped these
onto changes at the level of brain mechanisms. Sixth, we characterized the
effects of playful experience in the juvenile period on subsequent adult behavior, including motor performance, cognitive capacities, emotional stability,
and social competence. Seventh, we identified some of the brain mechanisms
that are altered by play experience and that impinge on various competencies.
From all these steps (and all are, to varying degrees, works in progress), we
began to develop a model to explain the many facets of play fighting. Notice
that this approach involves constructing a layer-cake that is capable of
explaining all aspects of play only when all the layers are in place. Beware of
short cuts: nature was not created to make life easy for human curiosity.

Our approach is one that, using a variety of perspectives, first deconstructs
a complex behavior and then reconstructs it. If the deconstruction is a reasonable reflection of the true content and rules of the behavior, then the
reconstruction should recreate a phenomenon that is close to the original.
There are several ways to conduct "analyses by synthesis".[1] We began with a
purely behavioral analysis, in which the components and their rules of interaction were characterized. If any piece of the behavioral performance remains
unexplained by the synthesis, then we know that something is missing in the
analysis. Only once we are confident that we understand the construction of
the behavior is it profitable to shift the level of the analysis so that the underlying endocrinological, physiological, and neural mechanisms can be characterized. Failure to characterize the behavior correctly can lead to poor models
of the underlying neurology. The importance of the initial behavioral analysis
in guiding neurological explanation is illustrated well in our experience of
studying the righting reflex.

When a cat falls, upside down, it will right itself in mid-air. To do this, the
cat seems to rotate its head and neck in one direction, and then rotates its
shoulders and then its pelvis in the same direction, until its whole body has
rotated 180°. The completion of this sequence results in the cat landing right

side up. The initial explanation for the underlying mechanism was something like this: visual and/or vestibular signals tell the cat that it is falling upside-down and these trigger the rotation of the head and neck toward the ground; in turn, the rotation of the neck stimulates receptors in the neck vertebrae that trigger the rotation of the shoulders; and this rotation involving the movement of the vertebrae of the mid-section stimulates the rotation of the lower body. Stimulation of one action creates the stimuli necessary for the next movement and so on, a classic chain reflex that produces a 180^0 rotation, starting at the head and neck, and ending at the tail. However, there are two problems with this explanation. First, if the cat's head and neck stabilize in space so that once prone, they remain there, the neck has to rotate in the opposite direction, to prevent the head and neck from being displaced away from facing the ground by the rotation of the shoulders. The same is true for the shoulders during rotation in the lower body. As the second neck rotation in the opposite direction does not doom the animal to make endless oscillatory rotations back and forth, it suggests that – at the very least – a modification of the traditional theory is required, to the effect that the neck reflex-stimulated shoulder rotation must be inhibited once the initial rotation has taken place. Second, when the head and neck rotate in one direction, there is a simultaneous rotation of the lower body in the opposite direction. That is, the tail end rotates before the stimuli created by vertebral movements can reach it! So, either the chain-reflex explanation needs to be modified, with both facilitative and inhibitory phases to explain the neck rotations, or a different explanation is needed to account for the concurrent rotation of the neck and tail – one that envisages the process not as a reflex, but as a motor program that is triggered in its entirety by the sensation of falling. In other words, an inadequate behavioral description can lead to an inaccurate physiological explanation. By the way, the counter-rotation by the tail end of the animal seems to stabilize the hindquarters, which thus anchors the rotation by the forequarters.[2]

In the past couple of decades, computerized models and moving robots have been used to great effect in testing whether a behavioral analysis has yielded components and rules of behavior that actually recreate the observed, behavioral output.[3] One of the best examples involves the huddling behavior of young rat pups. In their first week or so of life, rat pups are pink and hairless and are not, as yet, able to self-regulate their body temperature. Thus, when their mother leaves them, their body temperature drops. To compensate for this cooling, the pups huddle together, much like people do who are stranded in the cold (figure 8.1). After many years of studying this behavior, Jeff Alberts concluded that this seemingly complex huddling behavior emerges from the interaction of simple rules: that the pups attempt to maintain contact with a vertical surface (i.e. thigmotaxis), and that they are more attracted to warm surfaces than to cool ones. These rules seemed to be sufficient to account for the huddling of rat pups of up to seven days of age, but for slightly older pups, another rule seemed to be required. Up until the seventh day, a rat pup is insensitive to the activity of adjacent pups. However, after this age, it is influenced by their activity, and becomes active or inactive based on the activity of its littermates. This became the third rule for pups older than seven days. Behaviorally, these rules seemed sufficient to explain the complex patterns of huddling; or were they? Joined by Jeff Schank, Jeff Alberts took this question to its logical conclusion by using a superb analysis by synthesis.

Figure 8.1 Groups of huddling rat pups are shown at two ages: 5 days and 20 days. The older pups (top group), unlike the younger pups, have their eyes open and their bodies are furred. (From Alberts, J. R. (2005). Infancy. In I. Q. Whishaw & B. Kolb (eds), *The Behavior of the Laboratory Rat. A Handbook with Tests* (pp. 266–277). Oxford University Press: New York, NY. © 2005, Oxford University Press, reprinted with permission.)

They began their analysis of huddling behavior by making a computerized model to represent the pups as 'stick images', and programmed them to move according to the three rules. Sure enough, when this model was programmed with only the first two rules, the cluster of sticks mimicked the behavior of pups that were seven days old or younger. But when all three rules were employed in the model, the sticks mimicked the behavior of the older pups. Next, Jeff Schank constructed small, metal robots that moved on wheels. When these "robopups" were programmed with the first two rules to guide their behavior, they resembled the younger pups, and when all three rules were employed, they behaved like the older pups. Thus, using computerized simulations and robotics, they were able to demonstrate that the rules derived from the analysis of live pups were sufficient to recreate their behavior.[4] This testing showed that the rules initially abstracted from the actual behavior were not arbitrary descriptions of some facets of the phenomenon, but rather were realistic renditions of the underlying rules that generate complex behavior.

Although we are far from this type of sophisticated testing in our analyses of play fighting in rodents, by using a description of the proposed rules (e.g. body targets attacked and defended, preferred tactics of attack and defense in particular contexts) and testing them for their ability to account for species differences, age differences, and so on, we have some measure of confidence that the evolutionary scenario that we developed in chapter 6 is not simply a descriptive epiphenomenon.[5]

Where do We Go from Here?

Our own history of studying play reveals how elusive a completely satisfying explanation is for this puzzling, yet intriguing, behavior. In the mid-1970s, it seemed obvious to us that a plausible explanation for juvenile play fighting hinged on it being used as a means of refining combat skills. Indeed, we began studying rats because we could exercise more control over their developmental experiences and so test, more effectively, whether play fighting did, in fact, improve their combat skills as adults. However, our confidence in the explanatory power of this practice hypothesis quickly waned as it became clear that play fighting in rats was about sex, not aggression. Furthermore, the manner in which the tactics of attack and defense are used in play fighting in rats is a poor vehicle with which to rehearse the motor actions useful for sex, much less aggression.[6] For similar reasons, explanations for play such as it being essential for the refinement of cognitive or social skills were equally disappointing. Eventually, as our studies of play in rats, and those from many different laboratories around the world, began to demonstrate, deficiencies in the motor, cognitive, or social capacities of play-deprived rats arose from a more general deficit in their emotional reaction to events. A fearful and anxious animal is not one that is fully capable of bringing to bear, in any given situation, all its motor and cognitive skills. Thus, when play fighting, animals are not refining motor, cognitive, or social skills, but rather are learning how to calibrate and match their emotional reactions to an unpredictable world. Now, wait a moment, hasn't this explanation come full circle? It seems to be agreed by all that play experience in the juvenile period is about calibrating something, the question is what exactly is being calibrated – motor, cognitive, social, or emotional skills?

We hope that, having read chapter 4, you are convinced that, to a large extent, improvements in motor, cognitive, and social skills arise indirectly through play acting on the improvement of emotional skills, or more accurately, in refining the calibration of one's emotional responses to unexpected events in the world. Of course, we do acknowledge, and provide evidence for, the possibility that some motor, cognitive, and social skills are improved, directly, by the experience of play. Nonetheless, we consider that the primary avenue for the improvement of all skills is via emotional calibration. What does seem to be beyond argument is that for a number of species – rats, rhesus monkeys, and humans included – an impoverished experience of play fighting in the juvenile period leads to adults that have deficiencies in a variety of skills. What remains is to delineate the exact nature of the playful experiences that are necessary to produce improvements in all these skills and to pinpoint, precisely, the direct and indirect mechanisms by which they do so. However, if play is all about calibration, haven't we all simply been arguing about what exactly is being calibrated? Not quite: by now, you should be wary of a one-theory-fits-all kind of conclusion.

Imagine an orangutan moving through the trees. It uses its heavy body to bend thick branches or thin tree trunks in the direction in which it wishes to travel, and then, when it is close enough to it, reaches out with one of its long arms and grasps the next tree or branch in its path. An adult orangutan moving through the forest in this way seems to do so without expending any effort and is fluid in its motion. Climb a tree and try moving from branch to branch in this manner – it is not easy. Similarly, when young orangutans begin to move among the trees on their own, rather than being carried by their mothers, their movements are inefficient. Clearly, they need to practice these movements, and indeed, juvenile orangutans spend a large part of their day in non-utilitarian climbing or locomotor play.[7] Such play could be improving their climbing skills directly, or doing so indirectly, by giving the young orangutans confidence in their ability to climb without falling. That is, all this play could be contributing to dampening emotional responses to frightening situations rather than to improving motor skills. After all, it is reasonable for orangutans that are inexperienced climbers to be frightened because, as a species, they spend more time in trees than do either chimpanzees or gorillas and have up to three times more fractures of their long bones, suggesting more falls than these other apes.[8] Of course, the two are not mutually exclusive; improved motor coordination can improve one's confidence, which, in turn, can motivate one to try something a little more difficult and perhaps, more frightening.

So, among orangutans, play helps to improve the motor skills they need for locomotion, and does so, in part, by allowing the animals the opportunity to calibrate their movements so as to make their way through the trees effectively. As pointed out by Michael Simpson, a monkey jumping from one branch to another uses many muscles and joints, such as those of its feet to propel itself and those of its hands to grasp oncoming branches. Failure to achieve this goal could be due to an uncontrolled action by any number of muscles or joints. Play provides the opportunity for the animal to gain mastery over its anatomy. Jumping and leaping during play allows the animal to keep all things constant except for one set of muscles, so that by varying that one parameter, the animal can experience altering the force, say, that propels it from one branch to another. As play progresses, the animal should have more and more experience with varying most of the elements in its anatomical system and so have better control in making alterations to the series of actions involved.[9] Some of our personal observations are consistent with this calibration perspective.

When we filmed play fighting in juvenile patas monkeys,[10] we recorded instances of playful leaps. One such episode is particularly pertinent to this discussion on calibration. The three juveniles of the troop ran up a tree, one after another, to a horizontal branch, and then leapt, one at a time, towards a large, grassy patch in the enclosure. After landing, they ran back to the tree to repeat the performance. This sequence was repeated at least ten times by each monkey. In each jump, the monkeys altered the force and the height of their leaps, so that each successive one was slightly different. From a calibration perspective, this makes sense, as small, controlled variations are made with each subsequent leap. However, our description is incomplete. As they prepared to land, the monkeys did the exact opposite of what any sane animal would do. When hurtling towards the ground, an animal will stretch its limbs downwards, and, for animals with a flexible spine, curve its back dorsally. This "parachute reaction" prepares an animal for landing on the ground and does so in such a manner as to reduce the likelihood of risk to vital organs.[11] However, our monkeys did not do this. Instead, once they had reached their maximum height above ground, they raised their limbs laterally, thus adopting a spread-eagled posture, which caused them to land, full force, on their bellies. Landings were accompanied by sickening thuds and each monkey, before returning to the tree for a repeat performance, took a few moments to regain its composure. Although the variation in the force and height of the leaps was consistent with the hypothesis that such playful leaps are used to calibrate the motor system, the landings are not. If anything, these hard landings on their bellies would have taught the monkeys that having fun sometimes involves pain, or that pain can sometimes be fun. Thus, these landings are more consistent with the idea that play serves the function of calibrating emotional responses.

But calibration, be it of the motor or the emotional system, may be insufficient to explain all the actions performed during play. For example, while studying play fighting in a captive troop of Tonkean macaques, a type of Old World monkey that hales from Sulawesi, Indonesia, Christine Reinhart, a doctoral student from our laboratory, filmed some peculiar leaps in a pair of juveniles.[12] The two juveniles met along a path that was enclosed, like a bower, by overhanging tree branches. They both noticed one particular overhanging branch that was lower than the others and took turns leaping upward to grab at it. As the animals squatted, aimed, and then leapt upward, they reached out with one hand (figure 8.2, left), clearly targeting the branch. Several such leaps ensued. As in Simpson's model, and as illustrated in the jump portion of the patas monkey leaps, each successive leap that was carried out by the Tonkean macaques was only a small variation on the one preceding and resulted in small differences in the force and height of the jump. But then something changed. As their jumping progressed, each of the monkey's jumps became

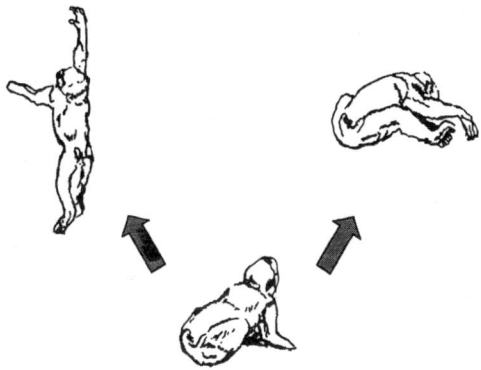

Figure 8.2 A Tonkean macaque is shown, crouching in preparation to leap into the air. In addition, two versions of the posture, when the monkey is at the apex of its leap, are shown. In the first version, its body is stretched out vertically, with one of its arms stretched upward as its hand reaches to grasp a branch. In the second version, the monkey's body is arched and nearly horizontal, with all four of its limbs stretched out laterally. (Drawing by Devin Cahoon, from a videotaped sequence courtesy of Christine Reinhart.)

more and more disconnected from what had seemed to be the original goal of their jumps – to reach for the branch. The monkeys added peculiar elements to each jump, such as spreading their hind legs and arching their backs (figure 8.2, right), which had nothing to do with successfully reaching the branch or in calibrating the movements composing a functional action.

It could be argued that these secondary, non-functional variations in the movements performed by these Tonkean macaques were designed for emotional calibration (chapter 4), or training for the unexpected.[13] However, for these latter explanations to be true, the expected variability in the actions occurring in the play should be designed to make achieving tasks more difficult, which would train the animals not to be flummoxed by unexpected impediments to their attempts to gain their desired goal. This is illustrated well in our juvenile rats, which stand on top of their supine partners with all four of their paws, thus reducing their control over their own movements and those of their partner, by making it harder to prevent their partners from successfully counterattacking and so causing a role reversal. The peculiar movements that the Tonkean macaques incorporated into their jumps made the attainment of their goal impossible, not just more difficult, and seemed to have been executed simply because they could be – perhaps for the sheer joy of being able to do them. Indeed, most definitions of play include some component that either explicitly or implicitly recognizes the intrinsically rewarding nature of play.[14]

That engagement in the performance of some behavioral actions can be self-rewarding has been recognized in the research traditions of psychology and animal behavior, and finding the underlying currency in the nervous system that drives such pleasure has been an important part of the tradition of behavioral neuroscience.[15] Perhaps, the mechanisms that make engaging in "behavior for its own sake" are being tapped during play? If this is so, then the challenge for the future will be to unravel what aspects of the playful performance can be explained by the functional benefits that such performance may bring from those aspects that are unleashed at the time of the performance itself by the neural and behavioral mechanisms engaged. After all, while the act of eating may be explained by the essential functional benefits that eating may have for survival and reproduction, not every individual act of eating can be so explained – our ever widening waistlines are a testimony to that fact. Our view is that more comprehensive theories of play will need to integrate the functions and the mechanisms that shape the playful performance more effectively.[16] As we have hopefully demonstrated in this book, such efforts at integration can be greatly facilitated by detailed knowledge about the development and evolution of play in useful lineages of animals, such as rodents.

Certainly, the example of the leaping monkeys shows that not all the actions performed during play can be resolved by any one theory that posits play experience as a means of calibration. Similarly, if play is self-rewarding, our current knowledge fails to explain why some actions should be performed while others are not. For the time being, then, these jumps are among the many actions that we see performed by playing animals for which we can do nothing more than shrug our shoulders and enjoy the performance.

ENDNOTES

Preface

1. Chudacoff, 2007; Sutton-Smith & Kelly-Byrne, 1984.
2. The utilitarian view of play can be seen in Plato's *Laws*, one of the earliest writings on the subject (1975). In modern times, this view is also evident in the theories of Maria Montessori (1967), which are still heavily endorsed by educators.
3. Power, 2000.
4. Hughes, 1999; Scarlett et al., 2005.
5. Groos, 1898.
6. Fagen, 1981; Smith, 1982.
7. Many of the chapters in Bekoff & Byers (1998) reflect this tradition.
8. Hole & Einon, 1984; Panksepp, Siviy, & Normansell, 1984; Thor & Holloway, 1984a.
9. Burghardt, 2001; Panksepp, 1998; Pellis & Pellis, 1998a, 2006; Siviy, 1998; Vanderschuren, Niesink & van Ree, 1997.
10. Of course, for those researchers, like ourselves, who have taken on a long-term approach to the study of play, these fluctuations in its interest have been felt directly. Over the past three decades, we have experienced comments ranging from "why in the world would you study play?" to "you study play, how cool is that?" The ease of publishing and of obtaining research funds also tends to follow these changes in attitudes.
11. Burghardt, 1984, 1988, 1998a, 2004, 2005.
12. In this book, by necessity of economy, we have had to gloss over the rich

history of thought on play that emerged, in the first half of the twentieth century, from students of psychology and animal behavior. For this, we apologize, and refer readers who wish to pursue this topic to texts that explicitly deal with this background. See the first two chapters in Burghardt (2005) and the chapter by Smith (1978) for thorough surveys of the ideas on play that emerged during the first half of the twentieth century. See also Sutton-Smith (1997) for a broad analysis of the various perspectives on play – from the humanities and social sciences as well as the natural sciences.

13. Pellis & Pellis, 2007.

Chapter 1

1. This description is based on our observations of a troop of Western lowland gorillas at the Calgary Zoo (Alberta, Canada). This particular episode depicts two three-year-old males.

2. In non-human animals, play is typically categorized as social, when it involves two or more partners; as object, when it involves an inanimate object; and as locomotor, when it involves movements not directed at a partner or object (Burghardt, 1998b). Although play may be similarly categorized in the human literature, it can also be characterized by different criteria. For example, human play can be classified on the basis of the degree of complexity in the cognitive manipulations involved; this leads to such categories as sensory-motor play, symbolic play and fantasy play (Smith, 2005). Given the emphasis on the description of behavior for non-human animals and the description of the psychological processes for humans, it is often difficult to compare the types of play identified in humans with that in other animals. Despite this problem, some correspondences can be clearly identified (Power, 2000). For our current purposes, play fighting in non-human animals and in humans seems to be comparable.

3. Bekoff & Byers (1981), Fagen (1981), and Smith (1978), all provide detailed reviews of these differing approaches to defining play. Miller (1973) shows how important our intuitions are in deciding whether a particular instance of behavior should be labeled as play.

4. Costabile et al.,1991; Smith et al., 1992; Smith, Smees, & Pellegrini, 2004.

5. See Burghardt (2005) for a detailed exposition of these five criteria. Each criterion shown can be fulfilled in more than one way, making it easier to compare across species. Comparison to the human literature (e.g. Hughes, 1999) shows that many of these criteria apply equally well to both non-human animals and to humans. For examples of play in what many people would consider exotic

animals, see Kramer & Burghardt (1998), Kuba et al. (2006), and Mather & Anderson (1999). Although the data were not collected with Burghardt's criteria in mind, even the play fighting-like behavior of immature cockroaches (Olomon, Breed, & Bell, 1976) seems a likely candidate for the play category. In support of this possibility is a recent report showing that another insect, the paper wasp, engages in interactions that also fit Burghardt's five criteria (Dapporto, Turillazzi, & Palagi, 2006).

6. Play typically uses up less than 20% of active time with an energetic cost usually less than 10% (Fagen, 1981; Martin, 1984a; Pellegrini, Horvat, & Huberty, 1998; Pellis, 1981a; Siviy & Atrens, 1992). In addition, many references can be found regarding injuries accrued and the risks of predation encountered during play (Fagen, 1981). One of the most compelling examples of these risks is a study of fur seals. Mortality among juvenile seals is highest when playing – which most often occurs in the water – due to predation by sharks (Harcourt, 1991a). However, the severity of the costs in terms of time, energy, and predation risk have been called into question (Bekoff & Byers, 1992; Caro, 1995; Martin, 1984b). A conservative approach posits that play has moderate costs, and so at best must provide moderate benefits (Martin & Caro, 1985). Another conclusion, and one consistent with the tenor of this book, is that we should be wary of making generalizations from studies of any one species (see endnote 10).

7. Over thirty different hypotheses have been proposed to explain the function of play (Baldwin, 1986; Fagen, 1981). Most are of the form that play in the juvenile phase is important for developing skills useful later in life. For example, play fighting is considered by some to be the means by which fighting skills are practiced in a non-dangerous context (Symons, 1978). Some are of the form that play provides some immediate benefits to the performer. For example, Barber (1991) suggests that play activity, of whatever form, is used by juveniles to regulate their nutritional loads. If the food consumed by juveniles is high in sugar relative to protein, play can burn off the excess sugar and so produce a more favorable balance of nutrients that are needed for growth. However, there is growing evidence that play can have both delayed and immediate benefits (Palagi, Cordoni, & Borgognini Tarli, 2004; Pellis, 2002a; Pellis, Pellis, & Foroud, 2005).

8. There is no doubt that infant animals are clumsy and so appear cute when they attempt to engage in some action, and it is this feature of immaturity that may be erroneously labeled as play (Fentress, 1983). In the juvenile period, there is a gradual unfolding of complex behaviors. Isolating the behaviors as they appear in the early phases of development from their appearance in later phases, may artificially create something that looks playful (Lazar & Beckhorn, 1974). More comprehensive criteria for defining play, such as those posited by Burghardt,

provide an empirical basis for deciding whether or not to label a particular case as play.

9. Evolution can be thought of as a "tinkering process" (Jacob, 1977), in which old parts are refurbished for new uses. See Futuyma (2005) for further details on dinosaur bones and feathers.

10. Burghardt (1984, 1988) developed the "surplus resources theory" (SRT) to characterize a set of features that, when brought together, increases the likelihood of play. For example, animals that regulate their body temperature endogenously, and have an extended juvenile phase and a large part of their diet provided by others, should be more likely to play than animals with the opposite pattern. Not surprisingly, mammals and birds have these properties and it is these animal groups that have produced the most playful members of the animal kingdom (Burghardt, 2005) (see Pellegrini, Dupuis, & Smith, 2006, for further elaboration of the SRT). Burghardt (2005) further proposes that additional processes may have then co-opted this initially functionless play into behavior that serves important functions. That such transformations may involve ever escalating degrees of change, is incorporated into Burghardt's model by identifying new levels of control and function, which are referred to as secondary and tertiary processes. As will become evident in the next few chapters, the data on rodents provide strong, comparative support for such a conception (Pellis, 1993).

11. Several laboratories around the world have contributed hundreds of studies on rats on the pattern and organization of play, its frequency and organizational changes during development, sex and strain differences, and the neural mechanisms that regulate such play. Several reviews give a flavor of the range of these studies (e.g. Panksepp, 1998; Pellis & Pellis, 1998a; Siviy, 1998; Vanderschuren et al., 1997). To a lesser extent, data are also available for a number of other laboratory rodents, such as Syrian golden hamsters (e.g. Cervantes et al., 2007; Delville et al., 2003) and voles (Pellis, Pellis, & Dewsbury, 1989; Pierce et al., 1991).

12. As will be detailed in later chapters, one way of gaining greater complexity in the form and function of play is for it to have more neural control mechanisms involved. One important avenue for these added controls is to have neural mechanisms in more recently evolved areas of the brain exert control over more ancestral brain mechanisms that are already regulating the performance of play (Pellis & Iwaniuk, 2004).

13. Play fighting in rats and some primates involves processes that ensure the experience of unpredictable movements – both their own movements and those of their partners. Such unpredictability, when experienced in early development, seems to be important for the development of an animal that is more capable of dealing with the vicissitudes of life (Pellis, Pellis, & Foroud, 2005: Pellis & Pellis,

2006). Play fighting that is able to generate unpredictable experiences involves novel control mechanisms in the cerebral cortex – the most evolved part of the mammalian brain (Foroud, Whishaw, & Pellis, 2004; Kamitakahara et al., 2007; Pellis, Pellis, & Whishaw 1992).

14. Whishaw & Kolb, 2004.

15. Another 25% of all mammal species are bats. This means that species of mammals with which most people are familiar (e.g. monkeys, dogs, sheep) come from a small number of orders (e.g. Primates, Carnivora, Artiodactyla) and represent only a minority of the species diversity in the class Mammalia (see Nowak, 1999).

16. See Murie & Michener (1984) for general information on the behavioral biology of ground squirrels. See Nunes et al. (1999) and Pasztor et al. (2001) for data on play in some ground squirrel species.

17. During domestication, animals that are the most docile, and hence the easiest to handle, are selected to breed. Over several generations of this selective breeding, a tame, domesticated version of the wild type is produced. However, selection for docility also results in the selection of correlated traits, such as early sexual maturity, which result in domesticated animals having more juvenile-like features (Trut, 1999).

18. See Apfelbach et al. (2005), Calhoun (1962), Lore & Flannelly (1977), and Blanchard & Blanchard (1990a) for social behavior and anti-predator behavior. See Prusky et al. (2002) and Whishaw et al. (2003) for perceptual and motor capacities. It is also worth pointing out that most of our work on rats has involved the Long Evans hooded strain, a strain that was first developed in 1915 when several white females from the Wistar Institute were bred with a wild male rat. Of the various comparisons alluded to above, the Long Evans strain appears to be most like the wild rat.

19. For reviews, see Burghardt (2001), Panksepp, Siviy, & Normansell (1984), Panksepp, 1998, Siviy (1998), Thor & Holloway (1984b), Vanderschuren et al., (1997).

20. Compared to the ease with which spoken language is acquired in the first decade of life, the many, many years of intensive instruction that it takes to become proficient at written language attests to the likelihood that even though spoken language is an evolved trait in our species, written language is a culturally variable epiphenomenon (Pinker, 1994).

21. As a general principle, a common approach to ascertain the adaptive value or function of biological traits is to use what are called "design features." For instance, if a fork is hypothesized to function as a tool for spearing food items, then it should have features that are well suited to that function – such as multiple, pointed tines. The more such design features are consistent with the proposed

function, then the greater the confidence in the likelihood that the proposed function is correct. Similarly, the greater the number of features of the system that are designed to bring about a particular outcome, the more likely that that product was evolved for that function rather than it being a by-product of other processes, such as a random confluence of mechanisms (for further discussion of this issue, see also Buss et al., 1998).

22. On the use of chemical treatments that block play while leaving exploratory loco-motion intact, see Pellis & McKenna (1995), Thor & Holloway (1983a), and Deak & Panksepp (2006). On anesthetizing the skin of a rat's nape, see Siviy & Panksepp (1987) and Lawrence et al. (2008).

23. Fagen, 1981.

24. Aldis, 1975; Pellis, 1988.

25. It should be noted that all experimentation with animals is carefully regulated and monitored by institutional animal welfare committees that adhere to strict national and international standards. These include ensuring that peer review has deemed the work of scientific merit and that the procedures used minimize any discomfort in the experimental animal.

26. Shakespeare (*Hamlet*, Act 1, Scene 5, line 166).

Chapter 2

1. Ewer, 1968,

2. Gervais & Wilson, 2005.

3. This has been shown by a series of studies from Jaak Panksepp's laboratory (e.g. Burgdorf & Panksepp, 2001; Panksepp & Burgdorf, 2000). However, there is debate about whether the chirping accompanying the tickling is best understood as analogous (independently evolved) or homologous (derived from a common ancestor) with human laughing (Gervais & Wilson, 2005; Panksepp, 2007; Panksepp & Burgdorf, 2003). Nonetheless, there is little doubt that rats find such tactile contact rewarding and that this characteristic chirping is an indication of the pleasure that they derive from the tickling – be it by a human experimenter or by another rat (Brunelli et al. 2006; Knutson, Burgdorf, & Panksepp, 1998, 1999).

4. Such social isolation leads to elevated frequencies of play fighting, but not to ele-vated levels of non-playful social behavior, such as social investigation (Panksepp, 1981; Panksepp & Beatty, 1980; Pellis & Pellis, 1990).

5. Pre-treatment with scopolamine, a drug that blocks the action of the neurotrans-mitter, acetylcholine, produces rats that actively explore the enclosure but are unresponsive to playful social contact (Deak & Panksepp, 2006; Pellis & McKenna, 1995; Thor & Holloway, 1983a).

6. Play fighting in rats activates both the dopamine and the opioid neurochemical systems of the brainstem, areas of the brain that include its pleasure centers (Panksepp, 1998; Siviy, 1998; Vanderschuren et al., 1997).

7. For repeated encounters with drugged partners, see Pellis & McKenna (1995) and for digging by anxious rats, see Pinel & Treit (1978).

8. Pellis, 2002a; Pellis, Pellis, & McKenna, 1993; Smith, Fantella, & Pellis, 1999.

9. Pellis & Pellis, 1987.

10. Pellis & Pellis, 1987; Pellis, Pellis, & McKenna, 1994; Pellis, Pellis, & Whishaw, 1992.

11. Pellis, 1989. To bring this point home, if we measure outcomes instead of movements, we may incorrectly conclude that the performances of two animals are the same, when, in fact, they are not. For example, rats can be trained to reach through a slot to grab and retrieve a piece of food. After a week or so of training, their success in retrieving food items can be well over 80%. Following damage to various areas of the brain's motor control mechanisms, the rats' performance can dwindle to below chance, but then, following further practice, their performance can return to pre-damage levels, and so appear recovered. Closer inspection of the rats' movements, however, show that they do not recover the same sequence of movements that they used prior to the damage, but rather, construct a new way of reaching. That is, they do not recover the lost reaching ability, but instead, develop a way of compensating for that loss. Only by examining the movements, rather than the outcomes, was this determined (Whishaw & Pellis, 1990; Whishaw et al., 1991). Such distinctions can be particularly important not only theoretically, but also practically, when applied to therapeutic interventions in people (Melvin et al., 2005; Whishaw et al., 2002).

12. Meaney & Stewart, 1981; Panksepp, 1981; Thor & Holloway, 1984b.

13. It is important to note that the developmental patterns that we describe here are based mostly on our own studies, using one particular strain of laboratory rat (Pellis & Pellis, 1990, 1997a; Pellis, 2002b). Different strains show variation in the absolute frequency of play fighting, the probability of defense and the likelihood of using different tactics of defense. For some strains, the sex differences are muted, or, for some aspects of play fighting, they are reversed. Nonetheless, the *relative* pattern is reasonably robust across most strains examined (Field et al., 2006; Pellis et al., 1997; Reinhart, Pellis, & McIntyre, 2004; Reinhart et al., 2006; Siviy, Baliko, & Bowers, 1997; Siviy et al., 2003). Furthermore, as we discussed in chapter 1, the behavioral pattern in pigmented strains, such as the Long Evans hooded, more closely reflects the behavior of wild rats than do the purely albino strains. Therefore, although it is not a perfect model for understanding play in rats, much less mammals in general, the basic story derived from the work on

Long Evans hooded rats provides a useful template for studying the evolution and function of play fighting. This is not unlike the variation seen among breeds of dog; all breeds share facets of the same behavior and developmental patterns seen in their wild progenitor, the wolf (Coppinger & Coppinger, 2001).

14. Aldis, 1975.
15. Pellis, 1988, 1993.
16. A common way to elicit serious fighting in rats (and many other rodents) is to place an unfamiliar, adult male into the home cage of a colony of rats (comprising both males and females). In this resident–intruder paradigm, the dominant, or alpha male, will attack the intruder. Most often, this resident directs bites at the intruder's flanks and lower dorsum. The intruder may retaliate by biting the side of the resident's face (Blanchard et al., 1977; Blanchard & Blanchard, 1990b). Higher intensity attacks by the resident may involve bites directed to the front of the face (Pellis, 1997). In some contexts where the stakes are likely higher, such as when a mother is protecting her pups, the attacker will most often go for the intruder's face (e.g. Kruk, van der Poel, de Vos-Frerichs, 1979; Sgofio et al., 1992).
17. Pellis & Pellis, 1998a.
18. For aggressive grooming during agonistic interactions, see Barnett (1975) and for nibbling during play fighting, see Pellis & Pellis (1998a).
19. Frank Beach (1976) was the first to note the active role of the female in regulating sexual encounters, but it was through a series of studies by Martha McClintock (1984) that the full richness of the female contribution was revealed. Nape nuzzling has been reported to occur when a male rat makes initial contact with a female preceding mounting (Calhoun, 1962; Pellis & Iwaniuk 2004; Whishaw & Kolb, 1985). The similarities in defensive tactics that are used to avoid nape contact during both playful and sexual encounters have been described in Pellis & Iwaniuk (2004). The general similarity of these tactics to those present in aggressive interactions can be found in Grant & MacIntosh (1963).
20. More detailed descriptions of the dynamics of attack and defense in rats, as it pertains to delivering bites and avoiding retaliatory bites, can be found in Blanchard et al. (1977) and Pellis & Pellis (1987).
21. We have also analyzed these dynamic complexities in several other species of rodents (e.g. Pellis & Pellis, 1992b; Pellis et al., 1992). All rat-like rodents seem to have the same offensive and defensive targets and so face the same constraints on their actions during fighting (Pellis, 1997).
22. Pellis & Pellis, 1987; Pellis & Iwaniuk, 2004.
23. Nunes et al., 1999, Pasztor et al., 2001.
24. Pellis & Pellis, 1992b.

25. During play fighting, animals have to straddle a fine line between competition and cooperation. If one animal persistently loses, then play fighting becomes a less attractive behavior. Essentially, animals need to curtail their competitiveness to ensure some degree of reciprocity (Altmann, 1962; Dugatkin & Bekoff, 2003; Pellis & Pellis, 1998b). Although the precise behavioral rules followed to achieve such reciprocity can vary across species, some means of ensuring fairness seem to be widespread across species that engage in play fighting (Pellis, Pellis, & Reinhart, in press).

26. Baum, 1984.

27. Although the majority of the literature on both humans and non-human animals supports the view that play fighting is male-typical (Ellis et al., 2008), there are some important caveats to bear in mind. First, there are exceptions to this view: there are species in which females play more than males (e.g. spotted hyena, Pederson et al., 1990; mantled howler monkeys, Zucker & Clarke, 1992) and also species in which there are no sex differences (e.g. cats, dogs and seals, Bekoff, 1974; Biben, 1983; Bradshaw, 1992; Caro, 1981; Harcourt, 1991b). Second, the traditional confounding of all forms of play fighting may have masked some significant sex differences. For example, ground squirrels engage in both aggressive and sexual play, but whereas males engage in more sexual play than do females, there are no sex differences in the frequency of aggressive play (Nunes et al., 1999; Pasztor et al., 2001).

28. See Pellis & Iwaniuk (1999b, 2000b) and references therein.

29. Males and females produce the same hormones, in that both sexes produce testosterone and estrogen. Thus, it is not the case that testosterone is a male hormone and estrogen a female hormone. Rather, what differentiates the sexes is either in the relative quantities of these hormones produced – with males producing more testosterone and females more estrogen – or in the sensitivity of the target tissue to these hormones (Daly & Wilson, 1983).

30. For a description of the organizational versus activational effects of gonadal hormones, see Baum (1984), and for the limitations in extending this framework to many, non-mammalian vertebrate groups, see Crews (1994).

31. Meaney, 1988; Meaney, Stewart, & Beatty, 1985; Pellis, Pellis, & McKenna, 1994; Smith, Forgie, & Pellis, 1998a; Thor & Holloway, 1983.

32. Smith et al., 1996; Smith, Forgie, & Pellis, 1998b.

33. Pellis, 2002b; Smith, Forgie, & Pellis, 1998a.

34. Field, Whishaw, & Pellis, 2004; Fitch & Dennenberg, 1998; Fitch, Cowell, & Dennenberg, 1998; Forgie & Stewart, 1994; Pellis, 2002b.

35. Laviola & Alleva, 1995; Terranova, Laviola, & Alleva, 1993; Terranova et al., 1998; van Oortmerssen, 1971; Walker & Byers, 1991.

36. Pellis & Pasztor, 1999; Wolff, 1981. These rudiments of social play in mice may still enable researchers to use modern genetic tools to dissect the mechanisms that make such play possible (Panksepp & Lahvis, 2007; Panksepp et al., 2007).

37. Although these perceptual differences between the sexes can account for some of the sex-specific frequencies in the use of different defensive tactics, a comparison between them in situations in which the attacker's orientation and the distance prior to the onset of the defensive action are the same, reveals that the large sex differences in age-related changes in types of defensive actions used are still robustly present (Pellis, Pellis, & McKenna, 1994; Pellis et al., 1997). Therefore, there must be non-sensory neural processes that regulate the majority of the sex differences in the age-related changes in playful defense, and so our arguments earlier in this chapter are still valid. Nonetheless, the differences in sensory bias (quick versus delayed response) need to be accounted for, and illustrate that there are many factors contributing to the production of complex outcomes.

38. Poole & Fish, 1975.

39. Pellis & Iwaniuk, 2004.

Chapter 3

1. It may be important for the reader to understand, in this context, our use of the term "control mechanism." What we have in mind is that if species X can only do A, but species Y can do both A and B, then to achieve B, species Y must have some control mechanism that is absent in species X. As will be argued later in the chapter, such a control mechanism can sometimes be linked to specific neural structures or circuits, and when this is known, the terms "control mechanism" and "neural mechanism" are synonymous. The exact nature of the relationship between the abstracted control mechanism and the neural mechanism that underlies it, and the way in which control mechanisms may interact with each other, involves a level of complexity that is beyond the scope of the present book. Nonetheless, readers may want to explore the hierarchical and feedback model developed by William Powers, which is probably one of the most sophisticated and useful approaches developed thus far for thinking about control. See Powers (1973) for the original exposition of this model, or Cziko (2000) for a less technical rendition.

2. Pellis & Iwaniuk, 1999a, 2004; Pellis & Pellis, 1998a.

3. Blanchard et al., 1977; Pellis & Pellis, 1987.

4. Pellis, 1997.

5. Pellis, 1988, 1993; Pellis & Pellis, 1998a.

6. Nowak, 1999.

7. Pellis & Iwaniuk, 2004.

8. Heymer, 1977; Miller, 1981.

9. This has been demonstrated for Richardson ground squirrels, in which the play fighting can be either sexual or aggressive (Pasztor et al., 2001), and for grasshopper mice, in which the play fighting can be constructed around either sexual or predatory targets (Pellis et al., 2000).

10. When there is a mixture of both, the exact proportion of the two can vary markedly even within closely related species. For example, for ground squirrels belonging to the same genus (*Spermophilus*), sexual play may comprise as much as 80% of all play fighting (Richardson's ground squirrel and Belding's ground squirrel), to as little as 20% (California ground squirrel) (Nunes et al., 1999; Pasztor et al., 2001; Pellis & Iwaniuk, 2004).

11. The method originated with a German scholar, Willi Hennig, but his original book was published in German in the 1950s, and so had little impact on international scholars. Even when it was published in English (1966), it did not have an immediate impact, but slowly, through the 1980s and 1990s, his approach came to dominate, and then largely displace, the alternatives (Brooks & McLennan, 1991, 2002; Harvey & Pagel, 1991).

12. See the introductory chapter in Lecointre & Le Guyader (2006) for a relatively easy-to-read explanation of the cladistic approach. For an elaboration of the example comparing salmon, lungfish and cows, see Gee (2001).

13. Maddison & Maddison (2000) provide one of the most commonly used programs and is the one that we have used in our own work.

14. For murid rodents, see Pellis (1988, 1993); for primates, see Mitchell (1979).

15. Pellis & Pellis, 1989; Pellis et al., 2000.

16. This is a core reason as to why Marc Bekoff has explored the role of play fighting in the evolution of empathy and morality in the animal kingdom (Allen & Bekoff, 2005; Bekoff, 2001). Of course, there has also been considerable thought and debate about the role of play in the development of morality during human childhood (for example, see Paglieri, 2005).

17. For further details on rats, see Pellis & Pellis, 1998b. A similar picture emerges for some monkeys. Some juvenile rhesus monkeys are so rough in their play that other juveniles in the troop quickly learn that it is not fun to play with them. These individuals usually end up being ostracized by other members of the playgroup (Suomi, 2005).

18. The 50:50 rule, first articulated by Stuart Altmann (1962), seems to fit the observations of many researchers (Fagen, 1981). As indicated, a 50:50 win-loss ratio is the most stable one in game theory simulations of play fighting (Dugatkin & Bekoff, 2003).

19. Thompson (1998) reports, for some rodents, a lack of restraint during play fighting. Other researchers have also reported that some animals can tolerate considerable deviation from parity (Bauer & Smuts, 2007). Researchers have also noted that for some species, the play fighting is so rough, that they hesitate to call it play, but rather, label it as quasi-aggression. A good example of this can be seen in the play fighting of juvenile pigs (Estes, 1993).

20. Detailed analyses of serious fighting in grasshopper mice show how this "lateral" maneuver can involve various orientations, including the attacker approaching the defender rump-first, which further reduces the capacity of the defender to retaliate (Pellis & Pellis, 1992b).

21. These principles of combat are fully explored in Blanchard & Blanchard (1994), Geist (1978), Pellis (1997), and Pellis & Pellis (1998b). In his studies of combat in mammals with horns and antlers, Valerius Geist nicely shows that when the ability of one animal to parry the thrusts of the other are curtailed, the opponent quickly delivers potentially lethal blows to the flanks (Geist, 1966). Indeed, in North American deer, 80% of mature males sport antler-induced scars on the flanks (Geist, 1986).

22. Pellis, 1981b; Pellis & Pellis, 1997a, 1998b.

23. Pellis, Pellis, & Reinhart, in press.

24. Clearly, much empirical work needs to be done to chart the full range of options available for how to keep play fights fair and so playful. Our own forays into this arena suggest that in several species of pigs there is no restraint in the actual delivery of attacks, but once the loser has signaled that the encounter should end, the winner does not press its advantage.

25. Pellis & Pellis, 1998a; Pierce et al., 1991.

26. Pellis & Pellis, 1988a, b; Pellis & Iwaniuk, 2004.

27. Kolb & Whishaw, 1996.

28. Bear in mind that this is a heuristic simplification that suits the present purpose, but note that in reality, such a simple division is becoming increasingly difficult to maintain. For example, evidence is accumulating that the cerebellum ("little brain"), a structure near the base of the brainstem, is intimately involved in higher order perceptual and cognitive processes (Bower & Parsons, 2003; Timmann et al., 2006).

29. For rats, see Panksepp et al. (1994) and Pellis, Pellis, & Whishaw (1992). For hamsters, see Murphy, MacLean, & Hamilton (1981). Although the study on decorticate Syrian golden hamsters was not detailed, given the distinctive predominance of supine defense and prolonged wrestling that characterizes play in this species, when the authors state that the play fighting of decorticated hamsters does not differ from that of intact controls, there is no reason to suspect otherwise.

30. Some of these circuits are better characterized than others, but the crude outlines of

the brain areas involved have emerged (Panksepp, 1998; Pellis et al., 1993; Siviy, 1998; Siviy et al., 1996; Trezza & Vanderschuren, 2008; Vanderschuren et al., 1997).

31. Pellis, Pellis, & Dewsbury, 1989; Pellis & Pellis, 1991b; Pellis & McKenna, 1992.

32. The conditioned place preference task is commonly used to assess motivation for certain tasks or items and has been applied to studying the neural circuitry of play motivation in rats (Siviy, 1998). Circuits coursing through the hypothalamus and connecting to the nucleus accumbens are particularly important in generating motivation to engage in particular kinds of behavior. This input to the nucleus accumbens involves the neurotransmitter dopamine, and constitutes a major part of the brain's reward system (Olds & Milner, 1954). It is the stimulation of this area, by increasing the release of dopamine, that leads to an increased desire to engage in a particular activity such as sex or eating. Unfortunately, stimulating this area by taking drugs, such as cocaine or heroin, leads to an increased desire to take the drugs and so sows the seeds of addiction (Robinson & Berridge, 1993).

33. Siegel, 2004.

34. The striatum, a subcomponent of the basal ganglia, is thought to be involved in the selection process of action patterns most appropriate for the task at hand (Hikosaka, 1998; Mink, 1996). Neonatal rats that are given intra-cerebral injections of 6-OHDA, a chemical that is toxic to dopamine-bearing neurons, and which disrupts the normal dopamine input to the striatum, leads to the interruption of this structure's normal function. When tested as juveniles, these rats still engage in high rates of playful attack and playful defense, but the cohesion of their defensive actions is severely curtailed (Pellis et al., 1993).

35. Aggleton, 2000; Baron-Cohen et al., 2000; Daenen et al., 2002; Jordan, 2003; Lewis & Barton, 2006; Maaswinkel et al., 1996; Meaney, Dodge, & Beatty, 1981; Pellis & Iwaniuk, 2002.

36. Adolphs, 1999; Bachevalier, Malkova, & Mishkin, 2001; Davis et al., 1995; Emery et al., 2001; Meunier et al., 1999; Pellis & Pellis, 1993; Prather et al., 2001; Suomi, 2005.

37. Pellis, Pellis, & Reinhart, in press.

38. Gallagher & Chiba, 1996; Ohman, 2002; Pellis & Iwaniuk, 2000b.

39. Bekoff, 1995; Pellis & Pellis, 1996.

40. Panksepp et al., 1994; Pellis, Pellis, & Whishaw, 1992.

41. Pellis, Pellis, & Whishaw, 1992.

42. Pellis & Pellis, 1991a; Pellis, Pellis, & McKenna, 1993; Smith, Fantella, & Pellis, 1999.

43. Pellis, Pellis, & Whishaw, 1992.

44. Kamitakahara et al., 2007; Pellis et al., 2006.

45. There are techniques that can record those areas of the brain that change in activity the most when engaged in a particular behavior. These techniques have revealed that in rats, play fighting leads to changes in the activity of brain areas ranging from the cortex to all the levels of the brainstem, with many of the areas most activated being some of the same as those that we discuss here (e.g. Burgdorf et al., 2006; Gordon et al., 2002; Siviy, 1998). Similarly, neurotransmitters, such as serotonin, that are widely distributed throughout the brain and connect both subcortical and cortical systems, have been shown to influence many different facets of play fighting in rats (e.g. Homberg et al., 2007).
46. Pellis & Iwaniuk, 2004.
47. Murphy, MacLean, & Hamilton, 1981.
48. Fagen, 1981; Fry, 2005.
49. Pellis & Pellis, 1993. Similar reluctance of subordinates to play with dominant partners has been reported in other species, such as coyotes (Bekoff, 1978).
50. Schmidt-Neilsen, 1984.
51. Many lineages of mammals have members in which the size of the brain has increased. Interestingly, the simple act of increasing size leads to changes in complexity. For example, by increasing the size of the brain, but not of the individual units – the neurons – communication between cells becomes more difficult due to the increasing distances between them. To solve this problem, larger brains tend to form local networks of neurons and then the networks communicate with those that are the nearest. Thus, with increasing brain size, there is a larger number of modular units – for example, the cortex of most small-brained mammals, such as rats, is divided into about twenty distinct areas, whereas, in larger-brained animals such as monkeys, there are up to fifty distinct cortical areas. Thus, greater organizational complexity is correlated with larger size, and larger brains grow more slowly. However, irrespective of size, depending on specific needs, some areas change independently of others. For example, in raccoons that use their hands to feel for food under water or beneath leaf litter, the area of the sensorimotor cortex devoted to the hands is expanded relative to other areas of the body. In contrast, in spider monkeys, which have a prehensile tail that they use like a fifth hand, the sensorimotor area devoted to the tail is expanded. Therefore, some changes in brain complexity are a by-product of changes in overall size (correlated evolution), but other changes involve alterations to meet specific species-typical needs and arise independently from other brain changes (mosaic evolution). In practice, both processes are probably involved in determining changes in brain structure. See Striedter (2005) for a detailed and accessible review of these issues.
52. McNamara, 1997.
53. Fagen, 1981.

54. Allman, 1999.

55. Iwaniuk, Nelson, & Pellis (2001) compared across fifteen different orders of mammals and species within three mammalian orders (primates, rodents, and marsupials). A comparative study of marsupial play by John Byers (1999) revealed a significant relationship between play prevalence and brain size, but when Iwaniuk and colleagues repeated the analysis, using a greater diversity of species, and a phylogenetic correction, the relationship disappeared. Similarly, in a comparative study of birds, it was found that play prevalence was not correlated with brain size (Diamond & Bond, 2003). What is meant by a phylogenetic correction? When species are compared, there is a potential problem. Let's say that twelve species are being contrasted, and ten out of the twelve fit the statistical pattern. It seems like a convincing relationship is present, but what if it turns out that those ten species are all very close relatives? It may be that the ancestor of those ten had a particular developmental and behavioral profile that was passed on to its ten descendent species. So, maybe we should count this cluster of species only once. Let's also assume that the two outlying species are distant relatives to each other and to the cluster of ten. With this corrected assessment, we now have only one of the three species showing the pattern – no longer very convincing. Thus, when comparing species of varying degrees of relatedness, it is potentially misleading to treat every species as a single datum. Several techniques are available to make a statistical correction on a sample of species for their relatedness. For example, the species are mapped onto a cladogram (see figure 3.5), and then the value for the compared variables are calculated at each node where the two species bifurcate. In the Independent Contrasts method, it is the values at the nodes that are graphed and statistically evaluated (Felsenstein, 1985). For an outline of the importance of Independent Contrasts in making cross-species neurobiological comparisons, see Iwaniuk, Pellis, & Whishaw (1999).

56. Lewis, 2000; Lewis & Barton, 2004.

57. Lewis & Barton, 2006.

58. Iwaniuk, Nelson, & Pellis, 2001; Pellis & Iwaniuk, 2002.

59. Pagel & Harvey, 1993.

60. For proponents of the view that an increased juvenile period evolved so as to provide greater opportunity to engage in play, see Groos (1898), Joffe (1997), Pereira & Altmann (1985). An alternative view is that an increased juvenile period evolved for other reasons (such as not squandering effort until resources become available), and that this led to an increased opportunity for play–as a by-product (Pagel & Harvey, 1993). Whether juvenility is caused by an increased need to play or whether increased juvenility produces greater opportunity to play, the expectation is that juvenility and play should be correlated.

61. Diamond & Bond, 2003; Pellis & Iwaniuk, 2000a; Ortega & Bekoff, 1987. It is also important to note that there may be important species differences in the exact age during the juvenile period that play may peak in occurrence. For some species, the bulk of the play may occur early in the juvenile period and then decline in the mid to late phases of this period (Fagen, 1993). Furthermore, different types of play may vary as to when they peak in occurrence during the juvenile period (Fairbanks, 2000). These caveats are likely to be important considerations for future, more detailed studies of the relationship between play and the juvenile period.

62. For our comparisons using primates and rodents, we used a phylogenetic correction – but with or without this correction, the relationship remained (Pellis & Iwaniuk, 2000a).

63. For the reasons that we give in endnote 51, it cannot be assumed that all the changes in the brain that can affect the complexity of play fighting necessarily arose for that purpose.

64. Brunelli & Hofer, 2007; Siviy, Baliko, & Bowers, 1997; Siviy et al., 2003.

65. The Fast and Slow rats were developed as an animal model for the study of epilepsy and their behavior and neural organization has been intensively studied (e.g. Anisman et al., 1997; Anisman & McIntyre, 2002; McIntyre et al., 2002; McIntyre et al., 1999; McIntyre, McLeod, & Anisman, 2004; McIntyre, Poulter & Gilby, 2002; Mohapel & McIntyre, 1998).

66. Corcoran & Teskey, 2004.

67. Reinhart, Pellis, & McIntyre, 2004; Reinhart et al., 2006.

Chapter 4

1. Baldwin (1986) provides a summary of some thirty proposed adaptive functions for play, with the majority assuming that the benefits are delayed. More recent reviews of both the human and non-human animal literature similarly emphasize the delayed benefits derived from playing during childhood (e.g. Bjorklund & Pellegrini, 2002; Power, 2000). Several authors have suggested that a primary benefit of playing is immediate, such as a means by which to shed excess calories (Barber, 1991). Adult play has been viewed in terms of its immediate benefits, such as for stress regulation (e.g. Darwish et al., 2001a, b; Palagi, Cordoni, & Borgognini Tarli, 2004; Palagi, Paoli, & Borgognini Tarli, 2006), the assertion of dominance (e.g. Mills, 1990; Smith, Fantella, & Pellis, 1999) and the maintenance of friendly relationships with group members (e.g. Palagi, 2006; Palagi & Paoli, 2007; Pellis & Pellis, 1992a).

2. C. L. Moore (1985) provides a comprehensive review of the majority of this literature, the bulk of which was published in the 1960s and 70s.

3. The importance of the active role of female rats during sexual encounters took longer to appreciate. Females not only solicit sexual approaches by using alluring signals such as wiggling their ears and darting out in front of the males, but they also regulate the pace at which mounts and copulations take place (Beach, 1976; McClintock & Adler, 1978). Needless to say, assessing female sexual behavior requires a larger floor space than that required for assessing male sexual performance. To our knowledge, there are no systematic studies available that have examined the effects of juvenile play fighting experience on adult, female sexual performance in rats. For this reason, most of our discussion involves what is known of the adult, male sexual performance.

4. Harper, 1972, 1973.

5. Goy & Goldfoot, 1974.

6. Aldis, 1975; Angier, 1995; Caro, 1988; Fagen, 1981; Miller & Byers, 1998; Pellis, 1981b; Smith, 1982; Symons, 1978.

7. Ellis et al., 2008. Of course, there are exceptions, and some of them support the view that play fighting practices fighting skills, but some do not. For example, spotted hyena females are more aggressive than males and are dominant over them, and also engage in more play fighting as juveniles (Pederson et al., 1990). However, female talapoin monkeys are dominant over males, yet as juveniles, males engage in more play fighting than do females (Wolfheim, 1977).

8. Hinde, 1975.

9. This approach is derived from the "argument from design", which originated as a theological one, whereby the detailed construction of the living world producing the goodness-of-fit between structure and function (e.g. bird wing and flight) was used to garner evidence for the existence of a supernatural creator. The development of the theory of natural selection by Darwin provided a materialistic mechanism with which to account for the diversity of living forms, and so, a non-supernatural means with which to explain the apparent goodness-of-fit (Ruse, 2003). Indeed, what Darwin's theory can explain better than the theological one is that on close inspection, the goodness-of-fit is sometimes not that good. For example, dead goldfish float upside down. Close examination of their behavior reveals that, when alive, in order for them to remain right side up, goldfish need to make continuous motions using their pectoral fins as sculls. The reason for this is that their swim bladder, which acts as ballast so that the fish can remain at a desired height in the water column, is near the ventral surface. With death, comes a cessation of sculling, and so, the inability of the body to remain right side up. If the swim bladder were placed deeper in the body, the fish would not need to scull with its fins to remain upright. This is a poor design that makes little sense if a supernatural creator were free to design organisms from scratch. However, it

makes more sense in a Darwinian world where present structures serving current functions need to be re-shaped from pre-existing structures – in the case of the fish swim bladder, it originated as a ventral outcropping of the esophagus with its initial function being to act as a rudimentary lung for fish living in shallow, oxygen-impoverished, fresh water (Romer, 1970).

10. Reeve & Sherman (1993) provide a strong critique of why trying to identify past selection on current function is not necessary to understand how traits currently contribute to survival and reproduction. This approach is typical of behavioral ecology in that it attempts to correlate an animal's behavioral traits with current ecological factors. However, behavioral ecologists have paid increasingly more attention to both behavioral and cognitive mechanisms by which individual animals act, thus placing more emphasis on the organization of traits and so past evolutionary influences (e.g. Krebs & Davies, 1997). Nonetheless, fields can differ in the emphasis they place on current and past influences of selection. For example, in evolutionary psychology, the focus of study is mostly on adaptations that have a clear evolutionary past (e.g. Buss, 2004), whereas, in Darwinian anthropology, the focus is on how current behavior may be adaptively modified to influence survival and reproduction under different environmental conditions (e.g. Low, 1999).

11. See Futuyma (2005) for further details on the origin of flight in birds. To capture the idea that the origin of a trait may not account for its current function, Gould & Vrba (1982) introduced the term "exaptation," where an exaptation is a trait that evolved for one purpose and now serves another.

12. The analysis of function from an evolutionary perspective is far more complex than anything that we could capture here. Readers are encouraged to explore some of the diverse literature for themselves. Good places to start are the collections of essays in Allen, Bekoff, & Lauder (1998) and Rose & Lauder (1996).

13. For the widow bird experiment and many other examples, see Andersson (1994). See also pp. 114–117 in Burghardt (2005) for a more detailed description of the different methodological approaches available to discern the adaptive value of behavior.

14. One last point needs to be made about function. Note that how we have used the term "function" here is synonymous with the term "adaptation." This is a common usage in disciplines that study behavior from a biological perspective. However, it can be confusing for those more accustomed to the literature in psychology or other fields of biology, such as functional morphology, where function refers to the immediate outcome of an action or structure. For example, the function of putting a piece of bread into a toaster is to toast the bread – no implication is made here as to whether this function confers adaptive value

(i.e. survival or reproductive advantage) to the performer (see Pellis (1997) and references therein). For this book, we will use the term function when the behavior in question confers survival and reproductive benefits, whether in the short term or long term. Wherever the two uses of function may be confounded, we will clarify our intended meaning.

15. Both the tactics of attack and defense that are typical of serious fighting are executed normally and seemingly effectively in rats that have been deprived of juvenile play fighting experience (Einon & Potegal, 1991; Potegal & Einon, 1989).

16. Groos (1898); (see chapter 3, endnote 60).

17. Field, Whishaw, & Pellis, 1996, 1997; Field, Watson, Whishaw & Pellis, 2005; Field & Pellis, 2008.

18. Pellis, Field, & Whishaw, 1999.

19. Martin & Caro, 1985.

20. Pellis & Iwaniuk, 2004. See also chapter 3.

21. Orgeur & Signoret, 1984.

22. Miller & Byers, 1998.

23. Estes, 1993; Rushen & Pajor, 1987.

24. This is based on our analyses of videotaped sequences of play fighting in several litters of Visayan warty pigs that we collected at the San Diego Zoo (2006, 2007).

25. Sharpe (2005a). Sharpe also tested other commonly espoused functional hypotheses about play fighting such as it being beneficial for social cohesion, but found them all wanting (Sharpe, 2005b, c; Sharpe & Cherry, 2003). However, it is important to note that Sharpe's data, which show the lack of an effect of play fighting experience on social cohesion, does not provide support for or against the hypothesis that we develop later in this chapter that play fighting functions to enhance social competence. Social cohesion may or may not be influenced by social competency.

26. Caro, 1980; Hill & Bekoff, 1977.

27. Power, 2000.

28. Deutsch & Larsson, 1974.

29. Bekoff, 1976; Burghardt, 1973.

30. Einon & Morgan, 1977; Einon, Morgan & Kibbler, 1978; Einon et al., 1981.

31. Holloway & Suter, 2004; Pellis et al., 1997; van den Berg, Hol et al., 1999.

32. Arakawa, 2002; 2007a; Einon & Morgan, 1977.

33. Einon & Potegal, 1991; Pellis, Field, & Whishaw, 1999; Potegal & Einon, 1989; Wright, Upton & Marden, 1991.

34. Holloway & Suter, 2004; Panksepp, 1981; Panksepp & Beatty, 1980; Pellis & Pellis, 1983, 1987, 1990; Varlinskaya & Spear, 2008.

35. Humphreys & Einon, 1981; Pellis & McKenna, 1995.

36. Bean & Lee, 1991; Hall, 1998; Robbins, Jones, & Wilkinson, 1996; Siviy, 1998.
37. Gottlieb, 2007.
38. In opposition to the commonly held view that it is high levels of testosterone that lead to aggression, in most cases, it is aggression that leads to high levels of testosterone. Whether the increased testosterone does, in turn, increase aggression, depends on the species, the sex, and the context. What is well established is that there is a feedback relationship between testosterone levels and aggression (Nelson, 2000).
39. For a review of the studies on the development of displays in gulls, see Groothuis (1993).
40. Pellis & McKenna, 1992; Rosenzweig & Bennett, 1972.
41. Similarly, if you have just returned home from a long drive and are quite tired, you may have difficulty in executing certain routines, such as unlocking the door, even if not being attacked by marauding birds! That performance is degraded by either hyper-arousal or hypo-arousal reflects the well-known Yerkes-Dodson law. This law posits that performance on some task follows an inverted U shape, with performance being best, or optimal, at the tip of the U, when arousal is at a moderate level (Mook, 1996).
42. Whishaw, Sarna, & Pellis, 1998.
43. Pellis, 2002a.
44. Metz, Schwab, & Welzl, 2001; Metz, Jadavji, & Smith, 2005; Smith & Metz, 2005.
45. McEwen & Sapolsky, 1995; Roozendaal, 2002.
46. Arakawa, 2002; Byrd & Briner, 1999; Einon & Potegal, 1991; Lore & Flannelly, 1977; van den Berg et al., 1999.
47. van den Berg et al., 1999; von Frijtag et al. 2002.
48. A problem with stress hormones such as corticosterone is that they are essential in helping the body deal with emergencies; shutting down unnecessary functions while channeling energy to the muscles that are needed to help the animal evade danger. Unfortunately, one of the non-essential functions that is shut down is the immune system. This is why it is important to lower levels of corticosterone quickly after an emergency is over. The longer corticosterone levels remain high, the more at risk an animal is of infections. Highly stressed animals are prone to ailments such as ulceration of the digestive tract (Sapolsky, 1995).
49. da Silva et al., 1996. Recent work has supported the view that in the juvenile period, a rat's experience with social play affects the development of its coping skills when it is faced with stressful social and non-social situations (Arakawa, 2007a, b).

50. Walf & Frye, 2007.

51. An anxiolytic is a substance that can reduce anxiety (such as the benzodiazepines, Valium®, Xanax®, etc.). In contrast, an anxiogenic is a substance that can induce a state of anxiety (such as sodium lactate, carbon dioxide, and caffeine).

52. Hol et al., 1999; Einon, Morgan, & Kibbler 1978; da Silva et al., 1996; van den Berg et al., 1999.

53. Arakawa, 2003.

54. There has been a growing awareness of the central role of emotional regulation in the effective execution of physical, cognitive (Damasio, 1994) and social skills. The latter have been shown to involve such things as being able to interpret social signals correctly (Lemerise & Arsenio, 2000). In addition, some authors have recognized the potential importance of play for emotional development (Bekoff, 2002; Fagen & Fagen, 2004; Kempes et al., 2008).

55. Biben, 1998.

56. Spinka, Newberry, & Bekoff, 2001.

57. For example, the opportunity to engage in play fighting has been shown to increase resiliency in brown bears (Fagen & Fagen, 2004) and children (Pellegrini et al., 2007).

58. Pellis, Pellis, & Dewsbury, 1989; Pellis, Pellis, & Reinhart, in press; Thompson, 1998.

59. Coles, 1996; Pellis & Pellis, 1997b.

60. Pellis & Pellis, 1987, 1990.

61. Foroud & Pellis, 2002, 2003; Pellis, Pellis, & Foroud, 2005.

62. Foroud & Pellis, 2002; Foroud, Whishaw, & Pellis, 2004.

63. Cortical modulation involves two different patterns of regulation for defense and offense. For defense, during the juvenile period, the cortex facilitates the use of the complete rotation tactic and inhibits the use of the partial rotation, whereas, for offense, in the pre-weanling and post-pubertal periods, the cortex inhibits the use of the unanchored position, but in the juvenile period, leaves it disinhibited (Foroud, Whishaw, & Pellis, 2004; Pellis, Pellis, & Whishaw, 1992). Even though altered in opposite directions, this pattern of modulation ensures the peculiar pattern of play fighting in the juvenile period.

64. See Kamitakahara et al. (2007) for data on the role of the motor cortex and Varlinskaya & Spear (2008) for data on the motivation by juvenile rats to engage in play fighting preferentially.

65. See Williams (1966, 1992) for an extended discussion on how to detect the effects of natural selection on features of organismic design.

66. We are employing the term "so-called," because it is expected that rats are born and raised in a world where they will have potential play partners. As this is generally true, a rat's developing neural and psychological mechanisms would

normally be exposed to play. Thus, raising a rat in a world where they do not receive any play experience is, in fact, abnormal, and so, the control condition can really be thought of as the experimental condition, since that condition imposes deprivation. This is a common problem in experimental research, in which the control condition is reduced to a state that no animal experiences in nature. As West & King (1987) have pointed out, organisms do not only inherit genes, but also, a species-typical, developmental niche (i.e. environmental context), and, for young rats, this niche includes having peers with whom to play.

67. Gruendal & Arnold, 1974.

68. See da Silva et al. (1996) for an example of how the lack of play experience leads to a later inhibition of exploration. The opposite has also been shown. Animals that have had greater opportunity for exploration and novelty seeking early in life are more likely to seek out novel experiences later in life (e.g. Neuringer, 2004). But there is a cost to be borne in mind. Rats are more susceptible to being taken by cats if they venture into an unfamiliar area (Elton, 1953). An experiment replicating this kind of finding was conducted using white-footed mice that were either familiar or unfamiliar with a testing room, which contained an owl as a predator (Metzgar, 1967). The mice were more likely to be taken by the owl when they ventured away from the walls and toward the center of the room. However, recent work has shown that under some circumstances, the best defense is to run towards a predator like an owl, rather than directly away (Eilam, 2005). This latter finding indicates that in some conditions, an animal that adopts a bolder strategy has a better chance of surviving (see also Dielenberg & McGregor, 1999).

69. Anisman et al., 1998; Boccia & Pedersen, 2001; Gandelman, 1992; Ogawa et al., 1994; Plotsky & Meaney, 1999; Pryce & Feldon, 2003; Pryce, Bettschen, & Feldon, 2001; Suchecki et al., 1995; van Oers et al., 1998. Note that the same principle appears to apply to human infants as well as to rats and mice (Allen, 1995; Zahr & Balian, 1995).

70. As noted in endnote 66, we must always be careful not to treat our experimental paradigms as if they were an uncontaminated reflection of nature. It is quite possible that under natural rearing conditions, the mothers of laboratory strains of rats and mice are not providing sufficient stimulation to their infants. That is, laboratory rodents that live in a world free of predators and social competitors, and where there is a constant abundance of food and water, may become slatternly mothers. The experimental treatments are likely supplementing what the infants are receiving from their mothers (Bateson & Martin, 2000). Studies in rats of naturally occurring variations in maternal licking and grooming show that such maternal attention is the mediating factor in an infant's response to being handled (Champagne et al., 2003). Although these experiments are important in

that they show that, when experienced in early infancy, moderate stress is important for the development of the stress-response system, what they do not show is that more stress early in life is better. Again, the normal developmental niche of rat and mice infants would be to experience moderate stress from being handled and separated from their mothers. It is the absence of such expected stress that is detrimental to infants! We would like to thank Terry DeVietti, who first brought the issue of what constitutes the control and experimental groups in these kinds of manipulations to our attention (personal communication, 1983).

71. Pellis & Pellis, 2006; Pellis, Pellis, & Foroud, 2005.

72. Pellis, 2002a; Whishaw et al., 2001.

73. Maternal separation and infantile handling of the intensity known to affect later stress-response functioning does not appear to affect the play fighting of juveniles, either in frequency or content (Arnold & Siviy, 2002), although these findings may be strain-dependent. Under some conditions, some strains may show an elevated level of play in the juvenile period, but do not appear to exhibit any changes in the content of their play (Siviy & Harrison, 2008). These data suggest that the play experience derived from the juvenile period adds to that derived from early infancy, rather than being dependent upon it.

74. Forgays & Forgays, 1952.

75. C. L. Moore, 1985; Pellis, Field, & Whishaw, 1999.

76. Gerall, 1963.

77. Pellis, Field, & Whishaw, 1999; Pellis et al., 2006.

78. The male rat can be physically placed, by the experimenter, onto the female in the correct orientation or the female can be pre-treated with haloperidol, a dopamine-blocker that renders it immobile, but able to maintain its erect posture. Decorticate rats and rats reared in isolation show comparable failures in executing successful mounting. Both are able to mount mobile females successfully after one correct mount, whether it was achieved with the help of the experimenter or not (C. L. Moore, 1985; Whishaw & Kolb, 1985).

79. Larsson, 1978.

80. Lore & Stipo-Flaherty, 1984; Pellis & Pellis, 1991a; Pellis, Pellis, & McKenna, 1993.

81. Pellis et al., 2006.

82. See Goldberg (2001) for a general and non-technical discussion of the frontal lobes as the executive brain. In rats, the prefrontal cortex matures during the juvenile period (Uylings et al., 1990). Recent experiments indicate that the volume and cell number of this area of the brain can be altered by experience with peer–peer interactions (Markham, Morris, & Juraska, 2007). Finally, experimental damage to two areas of the prefrontal cortex in rats, the OFC (Pellis et al.,

2006) and the mPFC (Bell, unpublished observations, 2007; Schneider & Koch, 2005), has been shown to affect some aspects of play fighting. Finally, for an example of imaging research that implicates the involvement of the prefrontal areas in the social interactions of humans, see Decety et al., 2004.

83. Kolb, 1995.

84. Bell, Kolb, & Pellis, 2007. It should be borne in mind that we have only recently documented the changes in the neurons of these prefrontal areas, and it is as yet unclear as to how these changes lead to altered function. The different direction of the effects on the neurons of the OFC and the mPFC may make sense in that one can imagine that, for some decisions, more information is better, but for others, more information may be distracting noise. If the number of possible synapses is taken as the amount of available information, then, in some neural systems, increasing the number of synapses may enhance performance, whereas in others, limiting the number of synapses to the bare essentials may be the route to enhancing performance. It is also important to note that earlier attempts to find changes in the brain following the manipulation of play experience in the juvenile period failed to do so. These earlier studies, however, focused on overall brain size or gross changes in brain areas (Renner & Rosenzweig, 1986). Our work, and that of others, has shown that the brain has to be examined at a finer level of analysis – at the level of the structure of cells (e.g. Bock et al., 2008) or the chemicals that they release (e.g. Gordon et al., 2003) – in order to reveal how play experience can influence brain development.

85. Kolb, 1990.

86. Pellis, Pellis, & Kolb, 1992. Recall the findings that were described earlier which showed that when reared with a non-playful peer, rats have an abnormal development of striatal neurons (endnote 36).

87. LeDoux, 1996.

88. Byers & Walker, 1995; Fairbanks, 2000.

89. Gordon et al., 2003.

90. Pellis & McKenna, 1992; Reinhart et al., 2006.

91. Bell et al., 2007.

Chapter 5

1. von Frijtag et al., 2002.

2. Arelis, 2006; Darwish et al., 2001a, b.

3. Romeo, Karatsoreos, & McEwen, 2006; Siviy, Harrison, & McGregor, 2006.

4. Burghardt, 2005.

5. Arelis, 2006. The effects of short-term isolation on increasing the frequency of

play have been demonstrated by many laboratories. In our laboratory, we routinely use twenty-four hours of social isolation to elevate the play levels of experimental subjects (e.g. Pellis & Pellis, 1990). See File & Vellucci (1978) for the early post-injection effects on anxiety of ACTH.

6. Fagen & Fagen, 2004; Henzi & Barrett, 1999; Merrick, 1977; Palagi, Cordoni, & Borgognini Tarli, 2004; Palagi, Paoli, & Borgognini Tarli, 2006; Schino et al., 1988.

7. Guard, Newman & Roberts, 2002; Panksepp, 1998; Niesink & van Ree, 1982.

8. Adam Miklosi, personal communication, 2007. However, it is worth noting that engaging in too much play can be unpleasant. If juvenile rats continue to engage in play for prolonged periods of time, not only does the frequency of play per unit time decrease, but the animals also shift from emitting vocalizations that denote pleasure to ones that denote displeasure or aversion (Burgdorf et al., 2006).

9. For a review on the roles of oxytocin and vasopressin in social bonding, see Lim & Young (2006). For some preliminary data on the role of these substances in the play of rats, see Panksepp (1998). Readers may be more familiar with the role of oxytocin in maternal behavior and lactation, and of vasopressin's role in the regulation of bodily fluids. However, these substances also play important roles in the regulation of social bonds beyond that of mother and infant (see the references cited above).

10. Pellis & Iwaniuk, 2004; Pellis & Pellis, 1998a.

11. Blanchard & Blanchard, 1990b.

12. Gheusi et al., 1994.

13. Pellis, 2002a; Pellis & Iwaniuk, 2004; Smith, Fantella, & Pellis, 1999.

14. Adams & Boice, 1983, 1989; Pellis, Pellis, & McKenna, 1993; Smith et al., 1998b.

15. Blanchard, Flannelly & Blanchard, 1988.

16. Pellis, Pellis, & McKenna, 1993. It should also be noted that in both semi-natural and free-living populations, in colonies of either laboratory rats or wild rats, researchers have reported such differences among subordinates. Of course, where there is more space, or, better still, when there is no barrier present, the wannabes keep well away from the dominant males. These wannabes are the most likely to leave the colony so as to establish their own, or to break into another existing colony. These are the beta subordinates: the ones that in our more limited enclosures play the role of the reluctant subordinate. In contrast, the gamma subordinates seem to prefer to remain as subordinates even when they have the opportunity to leave their colony. For details, see Barnett, 1975; Blanchard & Blanchard, 1990a; Calhoun, 1962; Lore & Flannelly, 1977.

17. Pellis, 2002a; Smith, Fantella, & Pellis, 1999. Again, it is important to note that

the available data on wild rats in free-living or semi-natural colonies indicate that these patterns of interaction in our laboratory rats are not artifacts, but rather, reflect the behavioral repertoire of wild rats living under naturalistic conditions (e.g. Berdoy, Smith, & MacDonald, 1995; Boreman & Price, 1972; Robitaille & Bovet, 1976; Seward, 1945).

18. Brueggeman, 1978; Pellis, 2002a.
19. Aldis, 1975; Fagen, 1981.
20. Pellis & Iwaniuk, 1999b.
21. Pellis & Iwaniuk, 2000b.
22. Pellis & Iwaniuk, 2004.
23. Palagi, Cordoni, & Borgognini Tarli, 2004; Palagi, Paoli, & Borgognini Tarli, 2006; Palagi & Paoli, 2007; Palagi, 2008, in press.
24. Bolles & Woods, 1964; Panksepp, 1998; Pellis, 2002a, b; Pellis & Pellis, 1998a; Siviy, 1998; Vanderschuren et al., 1997.
25. For a discussion of the need to integrate findings from different levels of analysis, see Thierry (2007).

Chapter 6

1. This is well illustrated in a study by Byers & Walker, 1995, which shows that there is a temporal overlap during development between the maturation of the cerebellum and the occurrence of the peak play period. From this, they hypothesize that the function of play is to shape the development of the cerebellum. Their data set included house mice and rats, but their proposed function does not explain why the social play in these two species is so different, and so consequently, it has no predictive power about how different species are expected to play, and certainly fails to explain why some closely related species do not play at all. Many other functional theories have the same limitations: data from different species, often from distantly related ones, involving different types of play, and differing complexities of the same type of play, are cherry-picked and packaged together. Such insensitivity to diversity renders even potentially good ideas weak in their construction and reduces their predictive power (e.g. Spinka et al., 2001).
2. Some definitions of play are so all-encompassing that they are only superficially useful. For example, labeling as play any behavior that is nonfunctional (e.g. Bekoff & Byers, 1981) leads to the play fighting of rats and cage stereotypies in zoo animals being lumped together. More sophisticated definitions, where a number of criteria have to be met before the play label is applied, fare much better, because an objective means is available by which to discriminate between play and cage stereotypies (e.g. Burghardt, 2005). Such a sophisticated definition would

certainly be able to label correctly the behavior of juvenile rats and mice as play. However, the profound differences in the play of these two species would be glossed over. Hence, although good definitions help us be sure that comparisons are appropriate – play is compared with play rather than with cage stereotypies – such definitions do not call for the detailed comparisons that are needed to evaluate the differences in the content of play. It is placing the diversity present in closely related species on a cladogram that gives one a real sense of the phenomenon.

3. Hogan, 2001.
4. Eilam & Golani, 1989; Golani & Fentress, 1985; Marler, 1987, 1991; V. C. Pellis, Pellis, & Teitelbaum, 1991; Pellis, Pellis, & Nelson, 1992.
5. Williams, 1992. One cost would be that the elements of the behavioral systems would not have the same degree of opportunity to be modified and integrated by experience. That is, a sudden and complete onset would greatly erode an animal's plasticity (capacity for change).
6. Coppinger & Smith, 1989; Gans, 1988; Hogan, 1988; Kortland, 1955.
7. Burghardt, 2005.
8. Eisenberg & Leyhausen, 1972; Hutson, 1975; Leyhausen, 1979; Ivanco, Pellis, & Whishaw, 1996; Pellis et al., 1988.
9. Gerall, 1963.
10. Jerry Hogan refers to such behavioral performance in the earlier developmental stages without any functional feedback as "prefunctional." That is, the behaviors appear independently of relevant, experience-based modification of those behaviors (Hogan, 1988, 2001). This is not to say that experiential feedback is not crucial for the development of the behaviors in question, only that feedback from their performance is unnecessary for their species-typical execution. Experience can have indirect and unpredictable effects on the development of later behavior. For example, when young squirrel monkeys are denied live insects as part of their diet, they will, as adults, fail to develop a fear of snakes (Masataka, 1994). But note that in this species, direct experience of snakes is not necessary to develop a fear of them.
11. West & King, 1987. See also Oyama (2000).
12. Bekoff & Fox, 1972.
13. It was Captain Cook, English maritime explorer *extraordinaire*, who ensured that ships were stocked with vitamin C-rich foods such as limes, which were able to last for long periods of time. This is the reason why the term "limeys" was used as a nickname for the British. See Anonymous. *Captain Cook (1728–1779)*. In Plantexplorers.com (Retrieved 7/15/07: http://www.plantexplorers.com/explorers/biographies/captain/captain-james-cook.htm).
14. Many groups of primates, including our ancestors, lost their capacity to

synthesize vitamin C (Jukes & King, 1975). Although some authors claim that cats may need supplemental amounts of vitamin C, it is clear that, unlike humans, a cat, as a carnivore, can manufacture most of what it needs. See Dunn, T. J., Jr. *Cats Are Different. Find Out How A Cat's Nutritional Needs Are? Different From A Dog's.* In Thepetcenter.com (Retrieved 7/25/2007: http://www. thepetcenter.com/imtop/catsaredif.html).

15. Among the best-studied examples of the loss of superfluous systems are the many cases in which animals have adapted to living in caves or in the ocean's depths. In such cases, their visual systems have degenerated. The process leading to such degeneration may include an absence of selection to weed out those individuals with the poorest vision. Furthermore, shutting down the production of an efficient visual system may be advantageous, as the resources that would otherwise be used can be shunted off into other systems that more effectively improve reproductive success (e.g. Jones & Culver, 1989).

16. Eibel-Eibesfeldt, 1961; Pinel et al., 1989; Riess, 1954.

17. Pellis, 1981a, b, 1983; Pellis & Pellis, 1982.

18. Einon et al., 1981; C. L. Moore, 1985; Pellis, Pellis, & Dewsbury, 1989.

19. Forgays & Forgays, 1952.

20. Burghardt, 2005; Martin & Caro, 1985; Thierry, 2005; West-Eberhardt, 2003.

21. Burghardt, 2005.

22. Bell, unpublished data; Pellis, Pellis, & Whishaw, 1992; Pellis et al., 2006; Kamitakahara et al., 2007; Schneider & Koch, 2005.

23. For examples for rats, see Hall (1998), Ladd et al. (2000), Ruedi-Betschen et al. (2005). For examples for primates, see Bryan & Riesen (1989), Dettling, Feldon, & Pryce (2002).

24. In contrast to birds, where paternal care is very common, it is rare in mammals for the father to contribute to the care and rearing of the young. However, some do exist. In these cases, it does appear that the father's interactions with the young contribute to their neural, physiological, and behavioral development – just as has been shown for the maternal contribution (e.g. Bredy et al. 2004; Dettling, Feldon, & Pryce, 2002; Pionanotti & Vieira, 2004; Young et al., 2001). However, it also appears that the core mammalian experience is for infants to receive critical stimulation from interactions with the mother, and then these may be added to or modified by paternal interactions, but more generally, following weaning, by peer–peer interactions (Champagne & Curley, 2005).

25. Biben & Suomi, 1993.

26. Harry Harlow raised infant monkeys with surrogate mothers that were models made either of bare wire or cloth-covered wire. Even though it was the bare wire mothers that provided them their milk (using cunningly inserted milk bottles

with nipples), the infants preferred remaining attached to their cloth-covered mothers, only going, periodically, to the wire model for a feed. Essentially, this work laid to rest the prevailing view that babies "learn" to love their "mothers" because they are the source of food – the "cupboard love" theory. In turn, these experiments revitalized studies of early patterns of attachment between infants and caregivers (usually the mother) in humans that were conducted by the likes of John Bowlby, Mary Ainsworth, and Mary Main. These studies resulted in changes in how infants were treated in neonatal wards, orphanages and the home. The history and the controversy surrounding Harlow's work are sensitively reviewed and analyzed in Blum (2002).

27. Anderson, Kenney, & Mason, 1977; Eastman & Mason, 1975; Mason & Berkson, 1975; Mason, 1978.

28. Capitanio, 1985.

29. Of course, although further experimental work is needed to determine the degree to which these mother–infant playful interactions add to the baseline level that is seen in rodents, it should be remembered that for these animals, most of the maternal handling effects are enacted in the first three weeks after birth, with their visual and auditory systems only coming on-line in the third week (Pryce & Feldon, 2003). Even though dependent on the mother, infant primates have their distance sensory receptors available from birth, and are able to engage in coordinated motor behavior earlier than rodents (Parker & McKinney, 1999), and so are better placed to take advantage of the experiences that are derived from mother–infant play. It should also be noted that, in primates, the data suggest that the deficits that occur in infants reared in social isolation can be ameliorated by exposing them as juveniles to other juveniles that have been socially reared. A means by which to ameliorate an isolate's condition is for a "therapist" monkey to coax them, gradually, into play (Harlow & Suomi, 1971). These findings support the idea that even though peer–peer interactions can build on that which is obtained in the mother–infant context, they are independent and can, to some degree, substitute for each other (see also Kempes et al., 2008).

30. For the experimental literature on the relationship between positive emotions and play, see Panksepp (1998). For how this may be used to improve animal welfare, see Boissy et al. (2007). For work on emotional development in children, see Denham et al. (1997) and Parke et al. (1992).

31. For data on primates, see Biben (1998), Levy (1979), Pellis & Pellis (1997a), Symons (1978), but it should also be pointed out that the same may be true for some other mammalian (Aldis, 1975; Pellis, 1984) and avian lineages (Diamond et al., 2006; Pozis-Francois, Zahavi & Zahavi, 1999; Pellis, 1981b).

32. Aldis (1975) was the first to notice the distinction between biting the

species-typical play target and mouthing a partner's body. That is, play fighting is a mixture of gaining an advantage over another by biting their play target and mouthing them. A detailed analysis of play between triads of otters clearly showed that the primary attacker directs its bites to the cheek, the species-typical play target, whereas the secondary attacker – that is, the third animal that joins in the fray – mouths and mostly chews on other parts of the body. The otter in the defensive position is the one that is usually targeted by the secondary attacker, and the body areas that it mouths are distant from the defender's mouth, and are, most commonly, the hind legs (Pellis, 1984). In sharp contrast, a comparable analysis of triadic play in rats shows that the secondary attacker lunges at the nearest nape available, which is typically that of the attacking member of the pair, the one standing on top of the supine defender. The secondary attacker moves passed other, easily accessible body parts in order to gain access to the nape. Thus, rats are target-fixated (Pellis & Pellis, 1998a) even though during social exploration, they will, like many murid rodents, nose most parts of their partner's body (Eisenberg & Kleiman, 1977). Therefore, the absence of generalized nosing during play fighting cannot be due to the absence of this behavior in the species' repertoire. Further support for target specificity in rats can be seen from the first appearance of play fighting, where the nape is the object of attack and defense (Pellis & Pellis, 1997b). In many primates and carnivores, generalized mouthing precedes fixating on the species-typical play target (Chalmers, 1984; Lazar & Beckhorn, 1974).

33. These episodes are based on our observations of a troop of Western lowland gorillas at the Calgary Zoo (Alberta, Canada) from 1993–2003. In total, we made detailed observations on seven juveniles (six males and one female). Most of the time, all seven followed the typical routine, but would then interject novel patterns into their play fights. Another example of a theme-breaker is when an attacker approaches and lunges at its partner while keeping its eyes closed at the same time! Clearly, this is not a tactic that is being used so as to enhance a successful attack. Although eye closing during play is present in monkeys, it is far more common in great apes (Russon, Vasey, & Gauthier, 2002).

34. See various chapters in Pellegrini & Smith (2005).

35. Hartwig, 2006; Nekaris & Bearder, 2006.

36. See Pellis & Iwaniuk (1999b, 2000b) and the papers cited therein.

37. Kleiman, 1983; Snyder et al., 2003; Wilson, 2005. To add one more tantalizing nugget to the panda story, our own observations of the play between a mother and her two-year-old son indicate that most of the play is initiated by the mother (San Diego Zoo, June, 2006). This further supports the functional argument that it is in the mother's self-interest to ensure that her offspring gain the necessary experiences through play.

38. For data on human mother and father play with infants and toddlers, see Paquette (2004), Paquette et al. (2003), Power & Parke (1983), Power (1985), Tamis-LeMonda et al. (2004). For a general review of how the father–child relationship may be considered in an evolutionary context, see Paquette (2004). For a critique and a reminder to take into account variations in this relationship across cultures, see Tamis-LeMonda, 2004. Our own observations of family groups of two species of gibbon (white-cheeked, Melbourne Zoo, Australia, 1998 and moloch, Howlett's Zoo, England, 2004) have shown that in these monogamous animals, most of the play fighting, especially when at its most rough, occurs between the father and the infant. In one case, the infant was female, and in the other, male, which suggests that both sexes may benefit from such rough housing.

39. Allman, 1999; Fagen, 1981.

40. Diamond & Bond, 2003; Iwaniuk, Nelson, & Pellis 2001.

41. The mouse can be thought of as containing essential stimuli that trigger predatory-like behavior in the cat. One such stimulus is its size – small – and another is that the mouse is a moving object. With age, more and more of these essential stimuli have to co-occur for the object to be recognized by the cat as a mouse (Baerends-van Roon & Baerends, 1979; Wolgin, 1982). Such elemental stimuli, in technical parlance, are termed "sign stimuli" and under the right conditions, can represent the whole stimulus package (Ewert, 2005). For example, if you know that you are walking through a field that contains snakes, a piece of straw being blown passed your foot by the wind will likely evoke a startle response.

42. Garvey, 1990.

43. See Burghardt (2005) for a detailed exploration of the conditions that make it possible for play to emerge from pre-play precursors.

44. Burghardt, 2005; Diamond & Bond, 2003; Power, 2000.

45. Pellis & Iwaniuk, 2004; Pellis & Pellis, 1983, 1987.

46. Byers, 1999; Diamond & Bond, 2003; Iwaniuk, Nelson, & Pellis 2001; Lewis, 2000.

47. See table 8.1, p.192 in Burghardt (2005). For values on brain size in many of those mammalian orders, see Iwaniuk, Nelson, & Pellis (2001).

48. Naturally, much better data for brain measurements and for the behavioral measures of play are needed to test these ideas. Again, the murid rodents illustrate the kind of data that are necessary for the behavioral side of the equation.

49. This is true for humans, several other species of primates, some hoofed mammals, and some carnivores. See Pellis & Iwaniuk (1999b, 2000b), Pellis (2002a), and Pellis, Pellis, & Foroud (2005) for extensive reviews of the literature from which this conclusion is drawn. Forms of play derived from completely different

behavioral contexts may have also converged toward the same function of social assessment. This is beautifully illustrated in a recent study on ravens (Bugnyar et al., 2007). These birds gauge whether an individual can be trusted in a situation involving competition for food by using information derived from their object play. That is, individuals that share during play can also be trusted not to steal one's hidden cache of food!

50. Biben, 1998; Pellis & Pellis, 1998b.
51. As pointed out by Sutton-Smith, 1997, play is ambiguous in many ways. If this is so, ambiguity may be a core feature of all play. In this instance, we specifically use the term "ambiguity" to represent the situation in play fighting where the intent of one's partner's actions is not always obvious. For example, one's partner may be cheating.

Chapter 7

1. The vertebrates are actually a subphylum of the larger phylum Chordata, a group that includes the lancets and sea squirts – animals with a dorsal nerve and dorsal notochord (a cartilaginous rod that stiffens the body), but not the segmented vertebrae of the Vertebrata (Lecointre & Guyader, 2006).
2. Burghardt, 2005.
3. Recall Burghardt's development of the surplus energy theory, which identifies a number of factors that generate play-like behavior, and then transforms that into play proper (see endnote 10 in chapter 1).
4. Brueggeman, 1978; Palagi, Cordoni, & Borgognini Tarli, 2004; Palagi, 2006; Palagi, Paoli, & Borgognini Tarli, 2006; Palagi & Paoli, 2007; Pellis & Iwaniuk, 1999b, 2000b; Pellis, 2002a.
5. Primates are divided into two main branches, the anthropoids (monkey-like primates) and the prosimians (pre-monkey-like primates). With the advent of cladistic taxonomy, a slightly different division has become evident – the strepsirrhines and the haplorrhines. These terms refer to the noses of the members of these respective groups. The strepsirrhines include all but one group of those previously included with the prosimians. Essentially, the strepsirrhines have a dog-like, wet nose, whereas the haplorrhines have a human-like, dry nose (Hartwig, 2006). One group of strepsirrhines, the lorisiforms, which includes pottos, lorises, and bush babies, are relatively small, nocturnal, tree-living animals with a diet that includes insects, and a social system with the least amount of social aggregation of all primates. It is the confluence of all these features that make the lorisiforms most like that which has been hypothesized to be typical of the ancestral primate condition (Nekaris & Bearder, 2006).

6. Anderson, 1970; Charles-Dominique, 1977; Charles-Dominique & Bearder, 1979; Epps, 1974.

7. Herrero & Hamer, 1977. See Pellis & Iwaniuk (1999b) for a review of more examples of courtship play in primates and other mammals.

8. Erhlich & Musicant, 1975; Erhlich, 1977; Evans & Poole, 1983, 1984; Geist, 1981, 1982; Newell, 1971; Paquette, 1994. Note that the loris and the bush baby belong to the same group of primates as the potto (see endnote 5).

9. Drea, Hawk, & Glickman, 1996; Erhlich & Musicant, 1975; Erhlich, 1977; Mills, 1990; Newell, 1971.

10. Ballard, Green & Granger, 2003; M. M. Moore, 1985, 1995; Ryan & Mohr, 2005.

11. Pellis, 2002a.

12. Dunbar, 1996.

13. Gervais & Wilson, 2005; Provine, 2000.

14. A large literature exists on humor, one that is too large to do justice to it here. However, we note some of that literature as it pertains to the key features that we have mentioned in the text and from which the interested reader can explore (Abel, 1998; Azim et al., 2005; Führ, 2002; Hay, 2000; Hobday-Kusch & McVittie, 2002; Priest & Swain, 2002).

15. See Corsaro (1997) on African-American banter. There is also a rich literature on the use of ridicule in "small scale societies," which appears designed to prevent any one individual from becoming too conceited (Boehm, 1999).

16. Jensvold & Fouts, 1993.

17. Pellegrini & Long, 2002; Pellegrini, 2003, 2006.

18. Sutton-Smith, 1982.

19. Fortunately, for Sergio, he moved, the next year, to a different school, where he remained for the rest of his secondary schooling. At this other, mixed-sex school, there was the occasional tough to deal with, but there did not exist that gang culture that had made the previous school so difficult.

20. Smith, Smees, & Pellegrini, 2004.

21. Of course, there are means by which the neural activity of animals can be monitored and these could be used to determine whether the brain's pleasure centers are activated in any particular bout of presumptive play (see Panksepp, 1998). Such methods are only feasible in highly constrained laboratory settings. Thus, for most instances of non-human animal play observed in the real world, we are not privy to the inner workings of the subjects' brains. Rather, we have to, as best as we can, make judgments about whether the behavior fit the criteria established for deciding whether a specific behavioral sequence is playful or not (see Burghardt, 2005). Nevertheless, experimental settings and measurements can be used that allow the animal to show us whether it finds the experience pleasurable. This can be achieved

either by giving it the opportunity to escape the test enclosure or by scoring whether it is evading its partner's playful overtures rather than promoting further contact (Reinhart et al., 2006; Varlinskaya, Spear & Spear, 1999).

22. Jones, 1983.
23. Conner, 1989; Pellegrini, 2003; Smith & Boulton, 1990.
24. Pellegrini, 1988, 1995a, b; Pellis & Pellis, 2007.
25. Bock, 2005; Fry, 2005; Pellegrini, 2006; Pellegrini & Smith, 1998; Smith, 2005.
26. Aldis, 1975; Blurton Jones, 1967, 1972; Fry, 1987.
27. Biben, 1998; Pellis & Pellis, 1998b; Pellis, Pellis, & Reinhart, in press.
28. Hughes, 1999; Clements & Fiorentino, 2004; Scarlett et al., 2005.
29. Fry, 2005; Power, 2000; Smith, 1997.
30. Bjorklund & Pellegrini, 2002; Freeman & Brown, 2004.
31. Pellegrini & Bjorklund, 1997.
32. Freeman & Brown, 2004.
33. Ikegami & Iizuka, 2007.
34. For sex differences in play and social competence, see Maccoby & Jacklin (1974) and Maccoby (1998). People with autism have impoverished social skills (White, Koenig & Scahill, 2007), and, perhaps not unrelated to the sex difference in these skills, there is a higher prevalence of autism in males. Indeed, it has been suggested that autism arises from the development of a hyper-masculine brain (Baron-Cohen, 2003).
35. Barrett, Dunbar, & Dunbar, 1992; Stone, 2008.
36. de Waal & Aureli, 2000; Fagen, 1981; Maccoby & Jacklin, 1974; Martin & Caro, 1985.
37. Kolb & Stewart, 1991.
38. Alcock, 2001.
39. Galdikas, 1981; MacKinnon, 1974; Mitani, 1985; Schurmann, 1982.
40. For examples of variation in the boldness-timidity gradient, see Fox (1972), Francis (1990), Lyons, Price, & Moberg (1988), Mendl & Paul (1990). For examples showing that bold or timid responses in one context are consistent with responses in other contexts, see Benus, Koolhaas, & van Oortmerssen (1987), Blanchard & Blanchard (1990a), Huntingford (1990). For examples that explore some of the neural and genetic differences that underlie these differences in temperament, see Benus et al. (1991), Clarke & Boinski (1995), Cools et al. (1990), Suomi (2005).
41. Blanchard et al., 1988; Pellis & McKenna, 1992.
42. Pellis & Pellis, 1990, 1991a, 1992a; Pellis, Pellis, & McKenna, 1993; Smith, Fantella, & Pellis, 1999.

43. Sutton-Smith & Kelly-Byrne, 1984.

44. For discussions of the role of sexual selection in humans in the differentiation of play fighting and aggression between the sexes, see Pellegrini & Archer (2005) and Pellegrini (2006).

45. Tremblay & Nagin, 2005.

46. Coie & Dodge, 1998. It is also worth pointing out that contrary to the practice so often applied in the past few decades of raising children's self-esteem in the belief that people are aggressive because of low self-esteem, the evidence shows that the opposite is true. Rather, people with high self-esteem, especially when it is not supported by actual ability, are the most likely to lash out violently when their self-image is challenged (Baumeister et al., 2005; Bushman & Baumeister, 1998).

47. Peterson & Flanders, 2005.

48. Burghardt & Burghardt (1972) describe such restraint in young animals.

49. Goldstein, 1995; Suomi, 2005; Wilmer, 1991.

50. Pellegrini, 1992.

51. Of course, while direct manipulation of these brain mechanisms is possible with rats and monkeys, it is not possible in humans. Nonetheless, judicious studies of people with neurodevelopmental disorders, such as autism, and the use of modern brain imaging technology, could provide the direct comparisons needed.

Chapter 8

1. Teitelbaum & Pellis, 1992.

2. Pellis, 1996. As an interesting aside, it worth noting that we initially noticed the counter-rotation of the lower body quite by accident. At the time, we were using rats, rather than cats, and after about 8 years of working with them, one of us, Sergio, had become allergic. During this particular experiment, Sergio was dropping the rats, upside down, in the air and Vivien was filming the trials using a high-speed camera. In order to obtain a "good" righting trial, the rat should be released once it relaxes in your hands, and to feel the rat's body more easily, it is better if one does not wear protective gloves. By the end of the session, Sergio had noticed welts on his wrists that were the result of being struck by the tails of the rats as they were righting. But how did they do that? This led us to look more closely at the video of their air righting, and, sure enough, as the rat was released, the head and neck began to turn one way and the lower body the other way, which resulted in a whip-like movement of its tail. If the rat's hindquarters had only begun to rotate *after* their forequarters had begun to rotate, Sergio would have been able to withdraw his hand in time and avoid being lashed by the tail (S. M. Pellis, Pellis, & Teitelbaum, 1991).

3. Braitenberg, 1987; R. A. Brooks, 1999; McFarland & Bösser, 1993. A maker of life-like robots may view this process of behavioral analysis as an "understanding by building" (Pfeifer & Bongard, 2007).

4. Alberts, 2007; May et al., 2006.

5. Our method of validation has hinged on the comparative approach, in which we first identified the key behavioral components and their rules of interaction via various descriptive techniques, then used the phylogenetic distribution of these elements across a set of related species, and finally, reconstructed the behavioral phenomenon and its distribution to see if the descriptive elements could account for all aspects of play fighting (Pellis, 2005).

6. Pellis & Pellis, 1998b.

7. Fagen, 1981.

8. van Schaik, 2004.

9. Simpson, 1976. It should also be noted that engagement in social play, including play fighting, has been shown, in children as well as in ground squirrels, to be associated with an improved motor performance in other contexts (Bar-Haim & Bart, 2006; Nunes et al., 2004a, b).

10. Pellis & Pellis, 1997a. Patas monkeys are Old World monkeys that live in Eastern Africa, and spend most of their time foraging on the ground. They have long limbs and the build of a greyhound, and, not surprisingly, can run very fast. Given their penchant for feeding out in the open grassland, their speed is their primary defense mechanism for escaping predators.

11. Monnier, 1970.

12. Christine Reinhart, at the time of writing, was comparing the organization of play fighting in two species of monkeys, the egalitarian and easy-going Tonkean macaque, and the despotic and highly-strung Japanese macaque. Differences in temperament between species offer another avenue for understanding species differences in the form of their play fighting and other social behavior (Thierry, Iwaniuk, & Pellis, 2000; Thierry, 2004). This is an avenue that is, as yet, poorly explored in rodents.

13. Spinka, Newberry, & Bekoff (2001) predict that play should be interspersed with variable actions so as to increase the experience of unexpected consequences. But it is unclear whether this prediction should include all kinds of variation in movement, or some subset. If all, then it is not possible to distinguish this theory from others that also predict that the movements that occur in play are variable. As more detailed empirical studies become available, greater precision in predicting the kinds of variation expected by one theory versus another become possible (e.g. Petrù et al., 2008).

14. Burghardt, 2005.

15. For the role of intrinsic reward in psychology, see Harlow (1953) and for animal behavior, see Hughes and Duncan (1988). For the mechanisms that underlie pleasure, see Cabanac (1992).

16. Indeed, as noted a number of times in this book, the adequate integration of function and mechanism is a major task that confronts all those interested in understanding behavior and its evolution (Thierry, 2005, 2007).

BIBLIOGRAPHY

Abel, M. H. (1998). Interaction of humor and gender in moderating stress outcomes. *Journal of Psychology*, **132**, 267–276.

Adams, N. & Boice, R. (1983). A longitudinal study of dominance in an outdoor colony of domestic rats. *Journal of Comparative Psychology*, **97**, 24–33.

Adams, N. & Boice, R. (1989). Development of dominance in rats in laboratory and seminatural environments. *Behavioral Processes*, **19**, 127–142.

Adolphs, R. (1999). The human amygdala and emotion. *Neuroscientist*, **5**, 125–137.

Aggleton, J. P. (Ed.). (2000). *The Amygdala: A Functional Analysis*. Oxford University Press: Oxford, UK.

Alberts, J. R. (2005). Infancy. In I. Q. Whishaw & B. Kolb (Eds.), *The Behavior of the Laboratory Rat. A Handbook with Tests* (pp. 266–277). Oxford University Press: New York, NY.

Alberts, J. R. (2007). Huddling by rat pups: Ontogeny of individual and group behavior. *Developmental Psychobiology*, **49**, 22–32.

Aldis, O. (1975). *Play Fighting*. Academic Press: New York, NY.

Alcock, J. (2001). *Animal Behavior. An Evolutionary Approach*. 7th Edn. Sinauer Associates: Sunderland, MA.

Allen, A.-M. (1995). Stressors to neonates in the neonatal unit. *Midwives*, **108**, 139–140.

Allen, C. & Bekoff, M. (2005). Animal play and the evolution of morality: An ethological approach. *Topoi*, **24**, 125–135.

Allen, C., Bekoff, M., & Lauder, G. (1998). *Nature's Purposes. Analysis of Function and Design in Biology*. MIT Press: Cambridge, MA.

Allman, J. M. (1999). *Evolving Brains.* Freeman Press: New York, NY.

Altmann, S. A. (1962). Social behavior of anthropoid primates: Analysis of recent concepts. In E. L. Bliss (Ed.), *Roots of Behavior* (pp. 277–285). Harper: New York, NY.

Anderson, C. O., Kenney, A. M., & Mason, W. A. (1977). Effects of maternal mobility, partner, and endocrine state on social responsiveness of adolescent rhesus monkeys. *Developmental Psychobiology,* 10, 421–434.

Anderson, M. (1970). A watched potto never grows: A chronicle of the prenatal and first months of a *Perodicticus potto. Discovery,* 6, 89–98.

Andersson, M. (1994). *Sexual Selection.* Princeton University Press: Princeton, NJ.

Angier, N. (1995). *The Beauty of the Beastly.* Houghton Mifflin: Boston, MA.

Anisman, H. & McIntyre, D. C. (2002). Conceptual, spatial, and cue learning in the Morris water maze in fast or slow kindling rats: Attention deficit comorbidity. *Journal of Neuroscience,* 22, 7809–7817.

Anisman, H., Lu, Z. W., Song, C., Kent, P., McIntyre, D. C., & Merali, Z. (1997). Influence of psychogenic and neurogenic stressors on endocrine and immune activity: Differential effects in Fast and Slow seizing rat strains. *Brain, Behavior & Immunity,* 11, 63–74.

Anisman, H., Zaharia, M. D., Meaney, M. J., & Merali, Z. (1998). Do early life events permanently alter behavioral and hormonal responses to stressors? *International Journal of Developmental Neuroscience,* 16, 149–164.

Anonymous. Captain Cook (1728–1779). In Plantexplorers.com (Retrieved 7/15/07: http://www.plantexplorers.com/explorers/biographies/captain/captain-james-cook.htm).

Apfelbach, R., Blanchard, D. C., Blanchard, R. J., Hayes, R. A., & McGregor, I. S. (2005). The effects of predator odors in mammalian prey species: A review of field and laboratory studies. *Neuroscience & Biobehavioral Reviews,* 29, 1123–1144.

Arakawa, H. (2002). The effects of age and isolation period on two phases of behavioral response to foot-shock in isolation-reared rats. *Developmental Psychobiology,* 41, 15–24.

Arakawa, H. (2003). The effects of isolation rearing on open-field in male rats depends on developmental stages. *Developmental Psychobiology,* 43, 11–19.

Arakawa, H. (2007a). Age-dependent change in exploratory behavior of male rats following exposure to threat stimulus: Effect of juvenile experience. *Developmental Psychobiology,* 49, 522–530.

Arakawa, H. (2007b). Ontogeny of sex differences in defensive burying behavior in rats: Effect of social isolation. *Aggressive Behavior,* 33, 38–47.

Arelis, C. L. (2006). *Stress and the Power of Play.* Unpublished M.Sc. thesis. Department of Neuroscience. University of Lethbridge: Lethbridge, AB, Canada.

Arnold, J. L. & Siviy, S. M. (2002). Effects of neonatal handling and maternal separation on rough-and-tumble play in the rat. *Developmental Psychobiology*, **41**, 205–215.

Azim, E., Mobbs, D., Jo, B., Menon, V., & Reiss, A. L. (2005). Sex differences in brain activation elicited by humor. *Proceedings of the National Academy of Sciences*, **102**, 16496–16501.

Bachevalier, J., Malkova, L., & Mishkin, M. (2001). Effects of selective neonatal temporal lobe lesions on socioemotional behavior in infant rhesus monkeys (*Macaca mulatta*). *Behavioral Neuroscience*, **115**, 545–559.

Baerends-van Roon, J. M. & Baerends, G. P. (1979). *The Morphogenesis of the Behaviour of the Domestic Cat, with Special Emphasis on the Development of Prey Catching*. North-Holland Publishing Co: Amsterdam, The Netherlands.

Baldwin, J. D. (1986). Behavior in infancy: Exploration and Play. In C. Mitchell & J. Erwin (Eds.), *Comparative Primate Biology*. Vol. 2A. *Behavior, Conservation, and Ecology* (pp. 295–326). Alan R. Liss: New York, NY.

Ballard, M. E., Green, S., & Granger, C. (2003). Affiliation, flirting, and fun: Mock aggressive behavior in college students. *The Psychological Record*, **53**, 33–49.

Barber, N. (1991). Play and energy regulation in mammals. *The Quarterly Review of Biology*, **66**, 129–147.

Bar-Haim, Y. & Bart, O. (2006). Motor function and social participation in kindergarten children. *Social Development*, **15**, 296–310.

Barnett, S. A. (1975). *The Rat: A Study in Behavior*. The University of Chicago Press: Chicago, IL.

Baron-Cohen, S. (2003). *The Essential Difference. The Truth about the Male and Female Brain*. Basic Books: New York, NY.

Baron-Cohen, S., Ring, H. A., Bullmore, E. T., Wheelwright, S., Ashwin, C., & Williams, S. C. R. (2000). The amygdala theory of autism. *Neuroscience & Biobehavioral Reviews*, **24**, 355–364.

Barrett, L., Dunbar, R. I. M., & Dunbar, P. (1992). Environmental influences on play behaviour in immature gelada baboons. *Animal Behaviour*, **44**, 111–115.

Bateson, P. & Martin, P. (2000). *Design for Life*. Simon & Schuster: New York, NY.

Bauer, E. B. & Smuts, B. B. (2007). Cooperation and competition during dyadic play in domestic dogs, *Canis familiaris*. *Animal Behaviour*, **73**, 489–499.

Baum, M. J. (1984). Hormonal control of sex differences in brain and behavior of mammals. In D. Crews (Ed.), *Psychobiology of Reproductive Behavior: An Evolutionary Perspective* (pp. 231–257). Prentice-Hall: Englewood Cliffs, NJ.

Baumeister, R. F., Campbell, J. D., Krueger, J. I., & Vohs, K. D. (2005). Exploding the self-esteem myth. *Scientific American*, **292**, 84–91.

Beach, F. A. (1976). Sexual attractivity, proceptivity, and receptivity in female mammals. *Hormones & Behaviour*, **7**, 105–133.

Bean, G. & Lee, T. (1991). Social isolation and cohabitation with haloperidol-treated partners: Effect on density of striatal dopamine D2 receptors in the developing rat brain. *Psychiatry Research*, **36**, 307–317.

Bekoff, M. (1974). Social play and play-soliciting by infant canids. *American Zoologist*, **14**, 323–340.

Bekoff, M. (1976). The social deprivation paradigm: Who's being deprived of what? *Developmental Psychobiology*, **9**, 499–500.

Bekoff, M. (1978). Social play: Structure, function, and the evolution of a cooperative social behavior. In G. M. Burghardt and M. Bekoff (Eds.), *The Development of Behavior: Comparative and Evolutionary Aspects* (pp. 367–383). Garland STPM Press: New York, NY.

Bekoff, M. (1995). Play signals as punctuation: The structure of social play in canids. *Behaviour*, **132**, 419–429.

Bekoff, M. (2001). The evolution of animal play, emotions, and social morality: On science, theology, spirituality, personhood, and love. *Zygon*, **36**, 615–655.

Bekoff, M. (2002). *Minding Animals*. Oxford University Press: New York, NY.

Bekoff, M. & Byers, J. A. (1981). A critical reanalysis of the ontogeny and phylogeny of mammalian social and locomotor play: An ethological hornet's nest. In K. Immelmann, G. W. Barlow, L. Petrinovich & M. Main (Eds.), *Behavioral Development: The Bielefeld Interdisciplinary Project* (pp. 296–337). Cambridge University Press: Cambridge, UK.

Bekoff, M. & Byers, J. A. (1992). Time, energy and play. *Animal Behaviour*, **44**, 981–982.

Bekoff, M. & Byers, J. A. (Eds.) (1998). *Animal Play: Evolutionary, Comparative, and Ecological Perspectives*. Cambridge University Press: Cambridge, UK.

Bekoff, M. & Fox, M. W. (1972). Postnatal neural ontogeny: Environment dependent or environment expectant? *Developmental Psychobiology*, **5**, 323–341.

Bell, H. C., Kolb, B., & Pellis, S. M. (2007). It's not child's play: Brain development is altered by horsing around. *Society for Neuroscience*, San Diego, CA, USA.

Benus, R. K., Koolhaas, J. M., & van Oortmerssen, G. A. (1987). Individual differences in behavioural reaction to a changing environment in mice and rats. *Behaviour*, **100**, 105–123.

Benus, R. K., Bohus, B., Koolhaas, J. M., & van Oortmerssen, G. A. (1991). Behavioral differences between artificially selected aggressive and non-aggressive mice: Response to apomorphine. *Behavioural Brain Research*, **43**, 203–208.

Berdoy, M., Smith, P., & MacDonald, D. W. (1995). Stability of social status in wild rats: Age and the role of settled dominance. *Behaviour*, **132**, 193–212.

Biben, M. (1983). Comparative ontogeny of social behavior in three species of South American canids: The maned-wolf, crab-eating fox, and bush dogs. Implications for sociality. *Animal Behaviour*, **31**, 814–826.

Biben, M. (1998). Squirrel monkey play fighting: Making the case for a cognitive training function for play. In M. Bekoff & J. A. Byers (Eds.), *Animal Play: Evolutionary, Comparative, and Ecological Perspectives* (pp. 161–2). Cambridge University Press: Cambridge, UK.

Biben, M. & Suomi, S. J. (1993). Lessons from primate play. In K. MacDonald (Ed.), *Parent-Child Play. Descriptions and Implications* (pp. 185–196). State University of New York Press: Albany, NY.

Bjorklund, D. F. & Pellegrini, A. D. (2002). *The Origins of Human Nature: Evolutionary Developmental Psychology.* American Psychological Association: Washington D.C., USA.

Blanchard, D. C. & Blanchard, R. J. (1990a). Behavioral correlates of chronic dominance-subordinance relationships of male rats in a seminatural situation. *Neuroscience & Biobehavioral Reviews,* **14,** 455–462.

Blanchard, D. C. & Blanchard, R. J. (1990b). The colony model of aggression and defense. In D. A. Dewsbury (Ed.), *Contemporary Issues in Comparative Psychology* (pp. 410–430). Sinauer Associates: Sunderland, MA.

Blanchard, R. J. & Blanchard, D. C. (1994). Environmental targets and sensorimotor systems in aggression and defense. In S. J. Cooper & C. A. Hendrie (Eds.), *Ethology and Psychopharmacology* (pp. 133–157). John Wiley & Sons: New York, NY.

Blanchard, R. J., Flannelly, K. J., & Blanchard, D. C. (1988). Life-span studies of dominance and aggression in established colonies of laboratory rats. *Physiology & Behavior,* **43,** 1–7.

Blanchard, R. J., Blanchard, D. C., Takahashi, T., & Kelley, M. J. (1977). Attack and defensive behaviour in the albino rat. *Animal Behaviour,* **25,** 622–634.

Blanchard, R. J., Hori, K., Tom, P., & Blanchard, D. C. (1988). Social dominance and individual aggression. *Aggressive Behavior,* **14,** 195–203.

Blum, D. (2002). *Love at Goon Park: Harry Harlow and the Science of Affection.* Perseus: Cambridge, MA.

Blurton Jones, N. G. (1967). An ethological study of some aspects of social behavior of children in nursery school. In D. Morris (Ed.), *Primate Ethology* (pp. 347–368). Weidenfeld & Nicolson: London, UK.

Blurton-Jones, N. G. (1972). Categories of child-child interaction. In N. G. Blurton-Jones (Ed.), *Ethological Studies of Child Behaviour* (pp. 97–127). Cambridge University Press: Cambridge, UK.

Boccia, M. L. & Pedersen, C. (2001). Animal models of critical and sensitive periods in social and emotional development. In D. B. Bailey & F. J. Symons (Eds.), *Critical Thinking about Critical Periods* (pp. 107–127). Brookes: Baltimore, MA.

Bock, J. (2005). Farming, foraging, and children's play in the Okavango Delta,

Botswana. In A. D. Pellegrini & P. K. Smith (Eds.), *The Nature of Play. Great Apes and Humans* (pp. 254–281). Guilford Press: New York, NY.

Bock, J., Murmu, R. P., Ferdman, N., Leshem, M., & Braun, K. (2008). Refinement of dendritic and synaptic networks in the rodent anterior cingulated and orbitofrontal cortex: Critical impact of early and late social experience. *Developmental Neurobiology*, **68**, 685–695.

Boehm, C. (1999). *Hierarchy in the Forest: The Evolution of Egalitarian Behavior*. Harvard University Press: Cambridge, MA.

Boissy, A., Manteuffel, G., Jensen, M. B., Moe, R. O., Spruijt, B., Keeling, L. J., Winckler, C., Forkman, B., Dimitrov, I., Langein, J., Bakken, M., Veisser, I., & Aubert, A. (2007). Assessment of positive emotions in animals to improve their welfare. *Physiology & Behavior*, **92**, 375–397.

Bolles, R. C. & Woods, P. J. (1964). The ontogeny of behavior in the albino rat. *Animal Behaviour*, **12**, 427–441.

Boreman, J. & Price, E. (1972). Social dominance in wild and domestic Norway rats (*Rattus norvegicus*). *Animal Behaviour*, **20**, 534–542.

Bower, J. M. & Parsons, L. M. (2003). Rethinking the "lesser brain." *Scientific American*, **289**, 50–57.

Bradshaw, J. W. S. (1992). *The Behaviour of the Domestic Cat*. C.A.B. International: Melksham, UK.

Braitenberg, V. (1987). *Vehicles: Experiments in Synthetic Psychology*. MIT Press: Cambridge, MA.

Bredy, T. W., Lee, A. W., Meaney, M. J., & Brown, R. E. (2004). Effect of neonatal handling and paternal care on offspring cognitive development in the monogamous California mouse (*Peromyscus californicus*). *Hormones & Behavior*, **46**, 30–38.

Brooks, D. R. & McLennan, D. A. (1991). *Phylogeny, Ecology and Behavior. A Research Program in Comparative Biology*. The University of Chicago Press: Chicago, IL.

Brooks, D. R. & McLennan, D. A. (2002). *The Nature of Diversity: An Evolutionary Voyage of Discovery*. The University of Chicago Press: Chicago, IL.

Brooks, R. A. (1999). *Cambrian Intelligence: The Early History of the New AI*. MIT Press: Cambridge, MA.

Brueggeman, J. A. (1978). The function of adult play in free-ranging *Macaca mulatta*. In E. O. Smith (Ed.), *Social Play in Primates* (pp. 169–192). Routledge: London, UK.

Bryan, G. K. & Riesen, A. H. (1989). Deprived somatosensory-motor experience in stumptailed monkey cortex: Dendritic spine density and dendritic branching of layer IIIB pyramidal cells. *Journal of Comparative Neurology*, **286**, 208–217.

Brunelli, S. A. & Hofer, M. A. (2007). Selective breeding for infant rat separation-induced ultrasonic vocalizations: Developmental precursors of passive and active coping styles. *Behavioural Brain Research*, **182**, 193–207.

Brunelli, S. A., Nie, R., Whipple, C., Winiger, V., Hofer, M. A., & Zimmerberg, B. (2006). The effects of selective breeding for infant ultrasonic vocalizations on play behavior in juvenile rats. *Physiology & Behavior*, **87**, 527–536.

Bugnyar, T., Schwab, C., Schloegl, C., Kotrschal, K., & Heinrich, B. (2007). Ravens judge competitors through experience with play caching. *Current Biology*, **17**, 1804–1808.

Burgdorf, J. & Panksepp, J. (2001). Tickling induces reward in adolescent rats. *Physiology & Behavior*, **72**, 167–173.

Burgdorf, J., Panksepp, J., Beinfeld, M. C., Kroes, R. A., & Moskal, J. R. (2006). Regional brain cholecystokinin changes as a function of rough-and-tumble play behavior in adolescent rats. *Peptides*, **27**, 172–177.

Burghardt, G. M. (1973). Instinct and innate behavior: Toward an ethological psychology. In J. A. Nevin & G. S. Reynolds (Eds.), *The Study of Behavior: Learning, Motivation, Emotion and Instinct* (pp. 322–400). Scott Foresman: Glenview, IL.

Burghardt, G. M. (1984). On the origins of play. In P. K. Smith (Ed.), *Play in Animals and Man*. Blackwell: Oxford, UK.

Burghardt, G. M. (1988). Precocial behavior, play, and the ectotherm-endotherm transition: Profound reorganization or superficial adaptation? In E. M. Blass (Ed.), *Handbook of Neurobiology*, Vol. 9, *Developmental Psychobiology and Behavioral Ecology*. Plenum Press: New York, NY.

Burghardt, G. M. (1998a). The evolutionary origins of play re-visited: Lessons from turtles. In M. Bekoff & J. A. Byers (Eds.), *Animal Play: Evolutionary, Comparative, and Ecological Perspectives* (pp. 1–26). Cambridge University Press: Cambridge, UK.

Burghardt, G. M. (1998b). Play. In G. Greenberg & M. Haraway (Eds.), *Comparative Psychology: A Handbook* (pp. 757–767). Garland: New York, NY.

Burghardt, G. M. (2001). Play: Attributes and neural substrates. In E. M. Blass (Ed.), *Handbook of Behavioral Neurobiology*, Vol. 13, *Developmental Psychobiology* (pp. 317–356). Kluwer Academic/Plenum Press: New York, NY.

Burghardt, G. M. (2004). Play: How evolution can explain the most mysterious behavior of all. In A. Moya & E. Font (Eds.), *Evolution: From Molecules to Ecosystems* (pp. 231–246). Oxford University Press: Oxford, UK.

Burghardt, G. M. (2005). *The Genesis of Play. Testing the Limits*. MIT Press: Cambridge, MA.

Burghardt, G. M. & Burghardt, L. S. (1972). Notes on the behavioral development of two female black bear cubs: The first eight months. In S. Herrero (Ed.),

Bears – Their Biology and Management, Vol. 23 (pp. 255–273). IUCN: Morges, Switzerland.

Bushman, B. J. & Baumeister, R. F. (1998). Threatened egotism, narcissism, self-esteem, and direct and displaced aggression. *Journal of Personality & Social Psychology,* **75,** 219–229.

Buss, D. M. (2004). *Evolutionary Psychology. The New Science of the Mind.* 2nd Edn. Allyn & Bacon: Boston, MA.

Buss, D. M., Haselton, M. G., Shackelford, T. K., Bleske, A., & Wakefield, J. C. (1998). Adaptations, exaptations, and spandrels. *American Psychologist,* **53,** 533–548.

Byers, J. A. (1999). The distribution of play behavior among Australian marsupials. *Journal of Zoology* (London), **247,** 349–356.

Byers, J. A. & Walker, C. (1995). Refining the motor training hypothesis for the evolution of play. *American Naturalist,* **146,** 25–41.

Byrd, K. R. & Briner, W. E. (1999). Fighting, nonagonistic social behavior, and exploration in isolation-reared rats. *Aggressive Behavior,* **25,** 211–223.

Cabanac, M. (1992). Pleasure: The common currency. *Journal of Theoretical Biology,* **155,** 173–200.

Calhoun, J. B. (1962). *The Ecology and Sociology of the Norway Rat.* US Public Health Service Publication, 1008. US Government Printing Office: Washington, D.C.

Capitanio, J. P. (1985). Early experience and social processes in rhesus macaques (*Macaca mulatta*): II. Complex social interaction. *Journal of Comparative Psychology,* **99,** 133–144.

Caro, T. M. (1980). The effects of experience on the predatory patterns of cats. *Behavioral & Neural Biology,* **29,** 1–28.

Caro, T.M. (1981). Sex differences in the termination of social play in cats. *Animal Behaviour,* **29,** 271–279.

Caro, T. M. (1988). Adaptive significance of play: Are we getting closer? *Trends in Ecology & Evolution,* **3,** 50–54.

Caro, T. (1995). Short-term costs and correlates of play in cheetahs. *Animal Behaviour,* **49,** 333–345.

Cervantes, M. C., Taravosh-Lahn, K., Wommack, J. C., & Delville, Y. (2007). Characterization of offensive responses during the maturation of play-fighting into aggression in male golden hamsters. *Developmental Psychobiology,* **49,** 87–97.

Chalmers, N. R. (1984). Social play in monkeys: Theories and data. In P. K. Smith (Ed.), *Play in Animals and Man* (pp. 119–141). Blackwell: Oxford, UK.

Champagne, F. A. & Curley, J. P. (2005). How social experiences influence the brain. *Current Opinions in Neurobiology,* **15,** 704–709.

Champagne, F. A., Francis, D. D., Mar, A., & Meaney, M. J., (2003). Variations in maternal care in the rat as a mediating influence for the effects of environment on development. *Physiology & Behavior*, **79**, 359–371.

Charles-Dominique, P. (1977). *Ecology and Behaviour of Nocturnal Prosimians. Prosimians of Equatorial West Africa.* Columbia University Press: New York, NY.

Charles-Dominique, P. & Bearder, S. K. (1979). Field studies of Lorisid behavior: Methodological aspects. In G. A. Doyle & R. D. Martin (Eds.), *The Study of Prosimian Behavior* (pp. 567–629). Academic Press: New York, NY.

Chudacoff, H. P. (2007). *Children at Play. An American History.* New York University Press: New York, NY.

Clarke, A. S. & Boinski, S. (1995). Temperament in nonhuman primates. *American Journal of Primatology*, **37**, 103–125.

Clements, R. L. & Fiorentino, L. (2004). *The Child's Right to Play. A Global Approach.* Praeger: Westport, CN.

Coie, J. D. & Dodge, K. A. (1998). Aggression and antisocial behavior. In W. Damon & N. Eisenberg (Eds.), *Handbook of Child Psychology*, Vol. 3 (pp. 779–862). Wiley: Toronto, ON.

Coles, B. L. K. (1996). *Neural Changes in Forelimb Cortex and Behavioural Development.* Unpublished M.Sc. Thesis. University of Lethbridge: Lethbridge, AB, Canada.

Cools, A. R., Brachten, R., Heeren, D., Willemen, A., & Ellenbroek, B. (1990). Search after neurobiological profile of individual-specific features of Wistar rats. *Brain Research Bulletin*, **24**, 49–69.

Conner, K. (1989). Aggression: Is in the eye of the beholder? *Play & Culture*, **2**, 213–217.

Coppinger, R. & Coppinger, L. (2001). *Dogs. A Startling New Understanding of Canine Origin, Behavior, and Evolution.* Scribner: New York, NY.

Coppinger, R. P. & Smith, C. K. (1989). A model for understanding the evolution of mammalian behavior. In H. Genoways (Ed.), *Current Mammalogy*, Vol. II (pp. 335–374). Plenum Press: New York, NY.

Corcoran, M. E. & Teskey, G. C. (2004). *Kindling: An Inquiry into Experimental Epilepsy.* Oxford University Press: Oxford, UK.

Corsaro, W. A. (1997). *The Sociology of Childhood.* Pine Forge Press: Thousand Oaks, CA.

Costabile, A., Smith, P. K., Matheson, L., Aston, J., Hunter, T., & Boulton, M. J. (1991). A cross-national comparison of how children distinguish serious and playful fighting. *Developmental Psychology*, **27**, 881–887.

Crews, D. (1994). Animal sexuality. *Scientific American*, **270**, 108–114.

Cziko, G. (2000). *The Things We Do: Using the Lessons of Bernard and Darwin to Understand the What, How and, Why of Our Behavior.* MIT Press: Cambridge, MA.

Daenen, E. W. P. M., Wolterink, G., Gerrits, M. A., & van Ree, J. M. (2002). The effects of neonatal lesions on the amygdala or ventral hippocampus on social behavior latter in life. *Behavioural Brain Research*, 136, 571–582.

Daly, M. & Wilson, M. (1983). *Sex, Evolution and Behavior*. 2nd Edn. Willard Grant: Boston, MA.

Damasio, A. R. (1994). *Descartes' Error*. Avon: New York, NY.

Dapporto, L., Turillazzi, S., & Palagi, E. (2006). Dominance interactions in young adult paper wasp (*Polistes dominulus*) foundresses: A playlike behavior? *Journal of Comparative Psychology*, 120, 394–400.

Darwish, M., Korányi, L., Nyakas, C., & Almeida, O. F. X. (2001a). Exposure to a novel stimulus reduces anxiety level in adult and aging rats. *Physiology & Behavior*, 72, 403–407.

Darwish, M., Korányi, L., Nyakas, C., & Ferenz, A. (2001b). Induced social interaction reduces corticosterone stress response to anxiety in adult and aging rats. *Klinikai és Kísérletes Laboratóriumi Medicina*, 28, 108–111.

da Silva, N. L., Ferreria, V. N. M., de Padua Gorabrez, A., & Morato, G. S. (1996). Individual housing from weaning modifies the performance of young rats on elevated plus-maze apparatus. *Physiology & Behavior*, 60, 1391–1396.

Davis, M., Campeau, S., Kim, M., & Falls, W. A. (1995). Neural systems of emotion: The amygdala's role in fear and anxiety. In J. L. McGaugh, N. M. Weinberger & G. Lynch (Eds.), *Brain and Memory: Modulation and Mediation of Neuroplasticity* (pp. 3–40). Oxford University Press: New York, NY.

Deak, T. & Panksepp, J. (2006). Play behavior in rats pretreated with scopolamine: Increased play solicitation by the non-injected partner. *Physiology & Behavior*, 87, 120–125.

Decety, J., Jackson, P. L., Sommerville, J. A., Chaminade, T., & Meltzoff, A. N. (2004). The neural bases of cooperation and competition: An fMRI investigation. *NeuroImage*, 23, 744–751.

Delville, Y., David, J. T., Taravosh-Lahn, K., & Wommack, J. C. (2003). Stress and the development of agonistic behavior in golden hamster. *Hormones & Behavior*, 44, 263–270.

Denham, S. A., Mitchell-Copeland, J., Strandberg, K., Auerbach, S., & Blair, K. (1997). Parental contributions to preschoolers' emotional competence: Direct and indirect effects. *Motivation & Emotion*, 21, 65–86.

Dettling, A. C., Feldon, J., & Pryce, C. R. (2002). Repeated parental deprivation in the infant common marmoset (*Callithrix jacchus*, Primates) and analysis of its effects on early development. *Biological Psychiatry*, 52, 1037–1046.

Deutsch, J. & Larsson, K. (1974). Model oriented sexual behavior in surrogate-reared rhesus monkeys. *Brain, Behavior & Evolution*, 9, 211–226.

de Waal, F. B. M. & Aureli, F. (2000). *Natural Conflict Resolution*. University of California Press: Berkeley, CA.

Diamond, J. & Bond, A. B. (2003). A comparative analysis of social play in birds. *Behaviour*, **140**, 1091–1115.

Diamond, J., Eason, D., Reid, C., & Bond, A.B. (2006). Social play in kakapo (*Strigops habroptilus*) with comparisons to kea (*Nestor notabilis*) and kaka (*Nestor meridionalis*). *Behaviour*, **143**, 1397–1423.

Dielenberg, R. A. & McGregor, I. S. (1999). Habituation of the hiding response to cat odor in rats (*Rattus norvegicus*). *Journal of Comparative Psychology*, **113**, 376–387.

Drea, C. M., Hawk, J. E., & Glickman, S. E. (1996). Aggression decreases as play emerges in infant spotted hyenas: Preparation for joining the clan. *Animal Behaviour*, **51**, 1323–1336.

Dugatkin, L. A. & Bekoff, M. (2003). Play and the evolution of fairness: A game theory model. *Behavioural Processes*, **60**, 209–214.

Dunbar, R. (1996). *Grooming, Gossip and the Evolution of Language*. Faber & Faber: London, UK.

Dunn, T. J., Jr. *Cats Are Different. Find Out How A Cat's Nutritional Needs Are Different From A Dog's*. In Thepetcenter.com (Retrieved 7/25/2007: http://www.thepetcenter.com/imtop/catsaredif.html).

Eastman, R. F. & Mason, W. A. (1975). Looking behavior in monkeys raised with mobile and stationary artificial mothers. *Developmental Psychobiology*, **8**, 213–222.

Eibel-Eibesfeldt, I. (1961). The interactions of unlearned behaviour patterns and learning in mammals. In J. F. Delafresnaye (Ed.), *Brain Mechanisms and Learning* (pp. 53–73). Blackwell: Oxford, UK.

Eilam, D. (2005). Die hard: A blend of freezing and fleeing as a dynamic defense – implications for the control of defensive behavior. *Neuroscience & Biobehavioral Reviews*, **29**, 1181–1191.

Eilam, D. & Golani, I. (1989). The ontogeny of exploratory behavior in the house rat (*Rattus rattus*): The mobility gradient. *Developmental Psychobiology*, **21**, 679–710.

Einon, D. F. & Morgan, M. J. (1977). A critical period for social isolation in the rat. *Developmental Psychobiology*, **10**, 123–132.

Einon, D. & Potegal, M. (1991). Enhanced defense in adult rats deprived of playfighting experience in juveniles. *Aggressive Behavior*, **17**, 27–40.

Einon, D. F., Morgan, M. J., & Kibbler, C. C. (1978). Brief periods of socialization and later behavior in the rat. *Developmental Psychobiology*, **11**, 213–225.

Einon, D. F., Humphreys, A. P., Chivers, S. M., Field, S., & Naylor, V. (1981). Isolation has permanent effects upon the behavior of the rat, but not the mouse, gerbil, or guinea pig. *Developmental Psychobiology*, **14**, 343–355.

Eisenberg, J. F. & Kleiman, D. G. (1977). Communication in lagomorphs and rodents. In T. A. Sebeok (Ed.), *How Animals Communicate* (pp. 634–654) Indiana University Press: Bloomington, IN.

Eisenberg, J. F. & Leyhausen, P. (1972). The phylogenesis of predatory behaviour in mammals. *Zeitschrift für Tierpsychologie*, **30**, 59–93.

Ellis, L., Hershberger, S., Field, E. F., Wersinger, S., Pellis, S. M., Geary, D., Palmer, C., Hoyenga, K., Hetsroni, A., & Karadi, K. (2008). *Sex Differences: Summarizing more than a Century of Scientific Research*. Psychology Press: New York, NY.

Elton, C. (1953). The use of cats in farm rat control. *British Journal of Animal Behaviour*, **1**, 151–155.

Emery, N., Capitanio, J. P., Mason, W. A., Machado, C. J., Mendoza, S. P., & Amaral, D. G. (2001). The effects of bilateral lesions of the amygdala on dyadic social interactions in rhesus monkeys (*Macaca mulatta*). *Behavioral Neuroscience*, **115**, 515–544.

Epps, J. (1974). Social interactions of *Perodicticus potto* kept in captivity in Kampala, Uganda. In R. D. Martin, G. A. Doyle & A. C. Walker (Eds.), *Prosimian Biology* (pp. 233–244). University of Pittsburgh Press: Pittsburgh, PA.

Erhlich, A. (1977). Social and individual behaviors in captive greater galago. *Behaviour*, **63**, 192–214.

Erhlich, A. & Musicant, A. (1975). Social and individual behaviors in captive slow lorises. *Behaviour*, **60**, 195–200.

Estes, R. D. (1993). *The Safari Companion. A Guide for Watching African Mammals*. Chelsea Green Publishing Company: Post Mills, VT.

Evans, S. & Poole, T. B. (1983). Pair-bond formation and breeding in common marmosets, *Callithrix jacchus jacchus*. *International Journal of Primatology*, **4**, 83–97.

Evans, S. & Poole, T. B. (1984). Long-term changes and maintenance of the pair bond in common marmosets, *Callithrix jacchus jacchus*. *Folia Primatologica*, **42**, 33–41.

Ewer, R. F. (1968). *Ethology of Mammals*. Plenum: New York, NY.

Ewert, J.-P. (2005). Stimulus perception. In J. J. Bolhuis & L-A. Giraldeau (Eds.), *The Behavior of Animals* (pp. 13–40). Blackwell: Malden, MA.

Fagen, R. A. (1981). *Animal Play Behavior*. Oxford University Press: New York, NY.

Fagen, R. A. (1993). Primate juveniles and primate play. In M. E. Pereira & L. A. Fairbanks (Eds.), *Juvenile Primates. Life History, Development, and Behavior* (pp. 182–196). Oxford University Press: New York, NY.

Fagen, R. A. & Fagen, J. (2004). Juvenile survival and benefits of play behaviour in brown bears, *Ursus arctos*. *Evolutionary Ecology Research*, **6**, 89–102.

Fairbanks, L. A. (2000). The developmental timing of primate play. A neural selection model. In S. T. Parker, J. Langer & M. L. McKinney (Eds.), *Biology, Brains, and*

Behavior. The Evolution of Human Development (pp. 131–158). School of American Research Press: Santa Fe, NM.

Felsenstein, J. (1985). Phylogenies and the comparative method. *American Naturalist,* 125, 1–15.

Fentress, J. C. (1983). A view of ontogeny. In J. F. Eisenberg & D. G. Kleiman (Eds.), *Advances in the Study of Mammalian Behavior* (pp. 24–64). Special Publication No. 7, The American Society of Mammalogists: Shippensberg, PA.

Field, E. F. & Pellis, S. M. (2008). The brain as the engine of sex differences in the organization of movement in rats. *Archives of Sexual Behavior,* 37, 30–42.

Field, E. F., Whishaw, I. Q., & Pellis, S. M. (1996). An analysis of sex differences in the movement patterns used during the food wrenching and dodging paradigm. *Journal of Comparative Psychology,* 110, 298–306.

Field, E. F., Whishaw, I. Q., & Pellis, S. M. (1997). The organization of sex-typical patterns of defense during food protection in the rat: The role of the opponent's sex. *Aggressive Behavior,* 23, 197–214.

Field, E. F., Whishaw, I. Q., & Pellis, S. M. (2004). Evidence for the necessity of pubertal ovarian hormones for the organization of female-typical patterns of dodging to protect a food item. *Behavioral Neuroscience,* 118, 1293–1304.

Field, E. F., Watson, N. V., Whishaw, I. Q., & Pellis, S. M. (2005). A masculinized skeletomusculature is not required for male-typical patterns of food-protective movements. *Hormones & Behavior,* 47, 49–55.

Field, E. F., Watson, N. V., Whishaw, I. Q., & Pellis, S. M. (2006). Play-fighting in androgen-insensitive tfm rats: Evidence that androgen receptors are critical for the development of adult playful defense but not playful attack. *Developmental Psychobiology,* 48, 111–120.

File, S. E. & Vellucci, S. V. (1978). Studies on the role of ACTH and of 5-HT in anxiety, using an animal model. *Journal of Pharmacy & Pharmacology,* 30, 105–110.

Fitch, R. H. & Dennenberg, V. H. (1998). A role for ovarian hormones in sexual differentiation of the brain. *Behavioral & Brain Sciences,* 21, 311–352.

Fitch, R. H., Cowell, P. E., & Dennenberg, V. H. (1998). The female phenotype: Nature's default? *Developmental Neuropsychology,* 14, 213–231.

Forgie, M. L. & Stewart, J. (1994). Effects of prepubertal ovariectomy on amphetamine-induced locomotor activity in adult female rats. *Hormones & Behavior,* 28, 241–260.

Forgays, D. G. & Forgays, J. W. (1952). The nature of the effect of free-environment experience in the rat. *Journal of Comparative & Physiological Psychology,* 45, 322–328.

Foroud, A. & Pellis, S. M. (2002). The development of "anchoring" in the play fighting of rats: Evidence for an adaptive age-reversal in the juvenile phase. *International Journal of Comparative Psychology*, 15, 11–20.

Foroud, A. & Pellis, S. M. (2003). The development of "roughness" in the play fighting of rats: A Laban Movement Analysis perspective. *Developmental Psychobiology*, 42, 35–43.

Foroud, A., Whishaw, I. Q., & Pellis, S. M. (2004). Experience and cortical control over the pubertal transition to rougher play fighting in rats. *Behavioural Brain Research*, 149, 69–76.

Fox, M. W. (1972). Socio-ecological implications of individual differences in wolf litters: A developmental and evolutionary perspective. *Behaviour*, 41, 298–313.

Francis, R. C. (1990). Temperament in fish: A longitudinal study of the development of individual differences in aggression and social rank in Midas cichlid. *Ethology*, 86, 311–325.

Freeman, N. K. & Brown, M. H. (2004). Reconceptualizing rough and tumble play: Ban the banning. *Advances in Early Education & Day Care*, 13, 219–234.

Fry, D. P. (1987). Differences between play fighting and serious fighting among Zapotec children. *Ethology & Sociobiology*, 8, 285–306.

Fry, D. P. (2005). Rough and tumble social play in humans. In A. D. Pellegrini & P. K. Smith (Eds.), *The Nature of Play* (pp. 54–85). Guilford Press: New York, NY.

Führ, M. (2002). Coping humor in early adolescence. *Humor*, 15, 283–304.

Futuyma, D. J. (2005). *Evolution*. Sinauer Associates: Sunderland, MA.

Galdikas, B. M. F. (1981). Orangutan reproduction in the wild. In C. Graham (Ed.), *Reproductive Biology of the Great Apes* (pp. 281–390). Academic Press: New York, NY.

Gallagher, M. & Chiba, A. A. (1996). The amygdala and emotion. *Current Opinion in Neurobiology*, 6, 221–227.

Gandelman, R. (1992). *The Psychobiology of Behavioral Development*. Oxford University Press: Oxford, UK.

Gans, C. (1988). Adaptation and the form-function relation. *American Zoologist*, 28, 681–697.

Garvey, C. (1990). *Play*. Harvard University Press: Cambridge, MA.

Gee, H. (2001). *Deep Time*. Fourth Estate: London, UK.

Geist, V. (1966). The evolution of horn-like organs. *Behaviour*, 27, 175–214.

Geist, V. (1978). On weapons, combat and ecology. In L. Krames, P. Pliner & T. Alloway (Eds.), *Advances in the Study of Communication and Affect*, Vol. 4, *Aggression, Dominance and Individual Spacing* (pp. 1–30). Plenum Press: New York, NY.

Geist, V. (1981). Behavior: Adaptive strategies in mule deer. In O. C. Wallmo (Ed.),

Mule and Black-tailed Deer of North America (pp. 157–223). University of Nebraska Press: Lincoln, NE.

Geist, V. (1982). Adaptive behavioral strategies. In J. W. Thomas & D. E. Toweill (Eds.), *Elk of North America* (pp. 214–277). Stackpole Books: Harrisburg, PA.

Geist, V. (1986). New evidence for high frequency of antler wounding in cervids. *Canadian Journal of Zoology*, **64**, 380–384.

Gerall, A. A. (1963). An exploratory study of the effects of social isolation on the sexual behaviour of guinea pigs. *Animal Behaviour*, **11**, 274–282.

Gervais, M. & Wilson, D. S. (2005). The evolution and functions of laughter and humor: A synthetic approach. *The Quarterly Review of Biology*, **80**, 395–430.

Gheusi, G., Bluthe, R., Goodall, G., & Dantzer, R. (1994). Social and individual recognition in rodents: Methodological aspects and neurobiological bases. *Behavioral Processes*, **33**, 59–88.

Golani, I. & Fentress, J. C. (1985). Early ontogeny of face grooming in mice. *Developmental Psychobiology*, **18**, 529–544.

Goldberg, E. (2001). *The Executive Brain. Frontal Lobes and the Civilized Mind.* Oxford University Press: New York, NY.

Goldstein, J. H. (1995). Aggressive toy play. In A. D. Pellegrini (Ed.), *The Future of Play Theory: A Multidisciplinary Inquiry into the Contribution of Brian Sutton-Smith* (pp. 127–147). State University of New York Press: Albany, NY.

Gordon, N. S., Kollack-Walker, S., Akil, H., & Panksepp, J. (2002). Expression of *c-fos* gene activation during rough and tumble play in juvenile rats. *Brain Research Bulletin*, **57**, 651–659.

Gordon, N. S., Burke, S., Akil, H., Watson, S. J., & Panksepp, J. (2003). Socially-induced brain "fertilization": Play promotes brain derived neurotrophic factor in the amygdala and dorsolateral frontal cortex in juvenile rats. *Neuroscience Letters*, **341**, 17–20.

Gottlieb, G. (2007). Probabilistic epigenesis. *Developmental Science*, **10**, 1–11.

Gould, S. J. & Vrba, E. S. (1982). Exaptation: A missing link in the science of form. *Paleobiology*, **8**, 4–15.

Goy, R. W. & Goldfoot, D. A. (1974). Experiential and hormonal factors influencing development of sexual behavior in the male rhesus monkey. In F. O. Schmitt & F. G. Worden (Eds.), *The Neurosciences: Third Study Program* (pp. 571–581). MIT Press: Cambridge, MA.

Grant, E. C. & McIntosh, J. H. (1963). A comparison of the social postures of some common laboratory rodents. *Behaviour*, **21**, 246–259.

Groos, K. (1898). *The Play of Animals.* Appleton: New York, NY.

Groothuis, T. G. G. (1993). Development of social displays: Form, development, form fixation and change in context. *Advances in the Study of Behavior*, **36**, 269–322.

Gruendal, A. D. & Arnold, W. J. (1974). Influence of preadolescent experiential factors on the development of sexual behavior in the albino rat. *Journal of Comparative & Physiological Psychology*, **86**, 172–178.

Guard, H. J., Newman, J. D., & Roberts, R. L. (2002). Morphine administration selectively facilitates social play in common marmosets. *Developmental Psychobiology*, **41**, 37–49.

Hall, F. S. (1998). Social deprivation of neonatal, adolescent, and adult rats has distinct neurochemical and behavioral consequences. *Critical Reviews in Neurobiology*, **12**, 129–162.

Harcourt, R. (1991a). Survivorship costs of play in the South American fur seal. *Animal Behaviour*, **42**, 509–511.

Harcourt, R. (1991b). The development of play in South American fur seals. *Ethology*, **88**, 191–202.

Harlow, H. F. (1953). Mice, monkeys, men and motives. *Psychological Review*, **60**, 23–32.

Harlow, H. F. & Suomi, S. J. (1971). Social recovery by isolation-reared monkeys. *Proceedings of the National Academy of Sciences*, USA, **68**, 1534–1538.

Harper, L. V. (1972). The transition from filial to reproductive function of coitus related responses in young guinea pigs. *Developmental Psychobiology*, **5**, 21–34.

Harper, L. V. (1973). Ontogeny of aberrant responses in the mating pattern of male guinea pigs. *Developmental Psychobiology*, **6**, 311–317.

Hartwig, W. (2006). Primate evolution. In C. J. Campbell, A. Fuentes, K. C. MacKinnon, M. Panger & S. K. Bearder (Eds.), *Primates in Perspective* (pp. 11–22). Oxford University Press: Oxford, UK.

Harvey, P. H. & Pagel, M. D. (1991). *The Comparative Method in Evolutionary Biology*. Oxford University Press: Oxford, UK.

Hay, J. (2000). Functions of humor in the conversations of men and women. *Journal of Pragmatics*, **32**, 709–742.

Hennig, W. (1966). *Phylogenetic Systematics*. University of Illinois Press: Urbana, IL.

Henzi, S. P. & Barrett, L. (1999). The value of grooming in female primates. *Primates*, **40**, 47–59.

Herrero, S. & Hamer, D. (1977). Courtship and copulation of a pair of grizzly bears, with comments on reproductive plasticity and strategy. *Journal of Mammalogy*, **58**, 441–444.

Heymer, A. (1977). *Ethological Dictionary*. Paul Parey: Berlin, Germany.

Hikosaka, O. (1998). Neural systems for control of voluntary action: A hypothesis. *Advances in Biophysics*, **35**, 81–102.

Hill, H. L. & Bekoff, M. (1977). The variability of some motor components of social play and agonistic behavior in infant eastern coyotes *Canis latrens*, var. *Animal Behaviour*, **25**, 907–909.

Hinde, R. A. (1975). The concept of function. In G. P. Baerends, C. Beer & A. Manning. (Eds.), *Function and Evolution in Behaviour* (pp. 3–15). Oxford University Press: Oxford, UK.

Hobday-Kusch, J. & McVittie, J. (2002). Just clowning around: Classroom perspectives on children's humor. *Canadian Journal of Education*, **27**, 195–210.

Hogan, J. A. (1988). Cause and function in the development of behavior systems. In E. M. Blass (Ed.), *Handbook of Behavioral Neurobiology*, Vol. 9 (pp. 63–109). Plenum Press: New York, NY.

Hogan, J. A. (2001). Development of behavior systems. In E. M. Blass (Ed.), *Handbook of Behavioral Neurobiology*. Vol. 13 (pp. 229–279). Kluwer Academic Publishers: New York, NY.

Hol, T., van den Berg, C. L., van Ree, J. M., & Spruijt, B. M. (1999). Isolation during the play period in infancy decreases adult social interactions in rats. *Behavioural Brain Research*, **100**, 91–97.

Hole, G. J. & Einon, D. F. (1984). Play in rodents. In P. K. Smith (Ed.), *Play in Animals and Children* (pp. 95–117). Blackwell: Oxford, UK.

Holloway, K. S. & Suter, R. B. (2004). Play deprivation without social isolation: Housing controls. *Developmental Psychobiology*, **44**, 58–67.

Homberg, J. R., Schiepers, O. J. G., Schoffelmeer, A. N. M., Cuppen, E., & Vanderschuren, L. J. M. J. (2007). Acute and constitutive increases in central serotonin levels reduce social play behaviour in peri-adolescent rats. *Psychopharmacology*, **195**, 175–182.

Hughes, B. O. & Duncan, I. J. H. (1988). The notion of ethological "need", models of motivation and animal welfare. *Animal Behaviour*, **36**, 1696–1707.

Hughes, F. P. (1999). *Children, Play and Development*. 3rd Edn. Allyn & Bacon: Needham Heights, MA.

Humphreys, A. P. & Einon, D. F. (1981). Play as a reinforcer for maze-learning in juvenile rats. *Animal Behaviour*, **29**, 259–270.

Huntingford, F. A. (1990). Fear and suppression of behavioural responses in fish. In P. F. Brain, S. Parmigiani, R. J. Blanchard, R. Mainardi & D. Mainardi (Eds.), *Fear and Defense* (pp. 41–67). Harwood Academic Publishers: London, UK.

Hutson, G. (1975). Sequences of prey-catching behaviour in the brush-tailed marsupial rat (*Dasyuroides byrnei*). *Zeitschrift für Tierpsychologie*, **39**, 129–150.

Ikegami, T. & Iizuka, H., (2007). Turn-taking interaction as a cooperative and co-creative process. *Infant Behavior & Development*, **30**, 278–288.

Ivanco, T. L., Pellis, S. M., & Whishaw, I. Q. (1996). Skilled movements in prey catching and in reaching by rats (*Rattus norvegicus*) and opossums (*Monodelphis domestica*): Relations to anatomical differences in motor systems. *Behavioural Brain Research*, 79, 163–182.

Iwaniuk, A. N., Nelson, J. E., & Pellis, S. M. (2001). Do big-brained animals play more? Comparative analyses of play and relative brain size in mammals. *Journal of Comparative Psychology*, 115, 29–41.

Iwaniuk, A. N., Pellis, S. M., & Whishaw, I. Q. (1999). Brain size is not correlated with forelimb dexterity in fissiped carnivores: A comparative test of the principle of proper mass. *Brain, Behavior & Evolution*, 54, 167–180.

Jacob, F. (1977). Evolution and tinkering. *Science*, 196, 1161–1166.

Jensvold, M. L. A. & Fouts, R. S. (1993). Imaginary play in chimpanzees (*Pan troglodytes*). *Human Evolution*, 8, 217–227.

Joffe, T. H. (1997). Social pressures have selected for an extended juvenile period in primates. *Journal of Human Evolution*, 32, 593–605.

Jones, C. B. (1983). Social organization of captive black howler monkeys (*Aloutta caraya*): Social competition and the use of non-damaging behavior. *Primates*, 24, 25–39.

Jones, R. & Culver, D. C. (1989). Evidence for selection on sensory structures in a cave population of *Gammarus minus* Say (Amphipoda). *Evolution*, 43, 688–693.

Jordan, R. (2003). Social play and autistic spectrum disorders. *Autism*, 7, 347–360.

Jukes, T. H. & King, J. L. (1975). Evolutionary loss of ascorbic acid synthesizing ability. *Journal of Human Evolution*, 4, 85–88.

Kamitakahara, H., Monfils, M.-H., Forgie, M. L., Kolb, B., & Pellis, S. M. (2007). The modulation of play fighting in rats: Role of the motor cortex. *Behavioral Neuroscience*, 121, 164–176.

Kempes, M. M., Gulickx, M. M. C., van Daalen, H. J. C., Louwerse, A. L., & Sterk, E. H. M. (2008). Social competence is reduced in socially deprived rhesus monkeys (*Macaca mulatta*). *Journal of Comparative Psychology*, 122, 62–67.

Kleiman, D. G. (1983). Ethology and reproduction of captive giant pandas (*Ailuropoda melanoleuca*). *Ethology*, 62, 1–46.

Kolb, B. (1990). Prefrontal cortex. In B. Kolb & R. C. Tees (Eds.), *The Cerebral Cortex of the Rat* (pp. 437–458). MIT Press: Cambridge, MA.

Kolb, B. (1995). *Brain Plasticity and Behavior*. Lawrence Erlbaum Associates: Mahwah, NJ.

Kolb, B. & Stewart, J. (1991). Sex-related differences in the dendritic branching of cells in the prefrontal cortex of rats. *Journal of Neuroendocrinology*, 3, 95–99.

Kolb, B. & Whishaw, I. Q. (1996). *Fundamentals of Human Neuropsychology*. 4th Ed. W. H. Freeman: New York, NY.

Kortland, A. (1955). Aspects and prospects of the concept of instinct (vicissitudes of the hierarchy theory). *Archives Néerlandaises Zoologie*, 11, 155–284.

Knutson, B. J., Burgdorf, J., & Panksepp, J. (1998). Anticipation of play elicits high-frequency ultrasonic vocalizations in young rats. *Journal of Comparative Psychology*, 112, 65–73.

Knutson, B. J., Burgdorf, J., & Panksepp, J. (1999). High-frequency ultrasonic vocalizations index conditioned pharmacological reward in rats. *Physiology & Behavior*, 66, 639–643.

Kramer, M. & Burghardt, G. M. (1998). Precocious courtship and play in emydid turtles. *Ethology*, 104, 38–56.

Krebs, J. R. & Davies, N. B. (1997). *Behavioural Ecology. An Evolutionary Approach.* 4th Edn. Blackwell: Oxford, UK.

Kruk, M. R., van der Poel, A. M., & de Vos-Frerichs, T. P. (1979). The induction of aggressive behavior by electrical stimulation in the hypothalamus of male rats. *Behaviour*, 70, 292–322.

Kuba, M. J., Byrne, R. A., Meisel, D. V., & Mather, J. A. (2006). When do octopuses play? Effects of repeated testing, object type, age, and food deprivation on object play in *Octopus vulgaris*. *Journal of Comparative Psychology*, 120, 184–190.

Ladd, C. O., Huot, R. L., Thrivikraman, K. V., Nemeroff, C. B., Meaney, M. J., & Plotsky, P. M. (2000). Long-term behavioral and neuroendocrine adaptations to adverse early experience. *Progress in Brain Research*, 122, 81–103.

Larsson, K. (1978). Experiential factors in the development of sexual behaviour. In J. B. Hutchinson (Ed.), *Biological Determinants of Sexual Behaviour* (pp. 55–86). John Wiley & Sons: New York, NY.

Laviola, G. & Alleva, E. (1995). Sibling effects on the behavior of infant mouse litters (*Mus domesticus*). *Journal of Comparative Psychology*, 109, 68–75.

Lawrence, R. C., Bonner, H. C., Newsom, R. J., & Kelly, S. J. (2008). Effects of alcohol exposure during development on play behavior and c-Fos expression in response to play behavior. *Behavioural Brain Research*, 188, 209–218.

Lazar, J. W. & Beckhorn, G. D. (1974). Social play or the development of social behavior in ferrets (*Mustela putorius*)? *American Zoologist*, 14, 405–414.

Lecointre, G. & Le Guyader, H. (2006). *The Tree of Life. A Phylogenetic Classification.* Harvard University Press: Cambridge, MA.

LeDoux, J. E. (1996). *The Emotional Brain.* Simon & Schuster: New York, NY.

Lemerise, E. A. & Arsenio, W. F. (2000). An integrated model of emotion processes and cognition in social information processing. *Child Development*, 71, 107–118.

Levy, J. S. (1979). *Play behavior and its decline during development in rhesus monkeys* (*Macaca mulatta*). Unpublished Ph.D. thesis. Department of Biology. University of Chicago: Chicago, IL.

Lewis, K. P. (2000). A comparative study of primate play behavior: Implications for the study of cognition. *Folia Primatologica*, 71, 417–421.

Lewis, K. P. & Barton, R. A. (2004). Playing for keeps: Evolutionary relationships between social play and the cerebellum in non-human primates. *Human Nature*, 15, 5–21.

Lewis, K. P. & Barton, R. A. (2006). Amygdala size and hypothalamus size predict social play frequency in nonhuman primates: A comparative analysis using independent contrasts. *Journal of Comparative Psychology*, 120, 31–37.

Leyhausen, P. (1979). *Cat Behavior. The Predatory and Social Behavior of Domestic and Wild Cats.* Garland STPM Press: New York, NY.

Lim, M. M. & Young, L. (2006). The neurobiology of social bonds and affiliation. In P. J. Marshall & N. A. Fox (Eds.), *The Development of Social Engagement: Neurobiological Perspectives* (pp. 171–196). Oxford University Press: Oxford, UK.

Lore, R. K. & Flannelly, K. (1977). Rat societies. *Scientific American*, 236, 106–118.

Lore, R. K. & Stipo-Flaherty, A. (1984). Postweaning social experience and adult aggression in rats. *Physiology & Behavior*, 33, 571–574.

Low, B. S. (1999). *Why Sex Matters. A Darwinian Look at Human Behavior.* Princeton University Press: Princeton, NJ.

Lyons, D. M., Price, E. O., & Moberg, G. P. (1988). Individual differences in temperament of domestic diary goats: Constancy and change. *Animal Behaviour*, 36, 1323–1333.

McClintock, M. K. (1984). Group mating in the domestic rat as a context for sexual selection – consequences for the analysis of sexual behavior and neuroendocrine responses. *Advances in the Study of Behavior*, 14, 1–50.

McClintock, M. K. & Adler, N. T. (1978). The role of the female during copulation in wild and domestic Norway rats (*Rattus norvegicus*). *Behaviour*, 67, 67–96.

McEwen, B. S. & Sapolsky, R. M. (1995). Stress and cognitive function. *Current Opinions in Neurobiology*, 5, 205–216.

McFarland, D. & Bösser, T. (1993). *Intelligent Behavior in Animals and Robots.* MIT Press: Cambridge, MA.

McIntyre, D. C., McLeod, W. S., & Anisman, H. (2004). Working and reference memory in seizure-prone and seizure-resistant rats: Impact of amygdala kindling. *Behavioral Neuroscience*, 118, 314–323.

McIntyre, D. C., Poulter, M. O., & Gilby, K. (2002). Kindling: Some old and some new. *Epilepsy Research*, 50, 79–92.

McIntyre, D. C., Kent, P., Hayley, S., Merali, Z., & Anisman, H. (1999). Influence of psychogenic and neurogenic stressors on neuroendocrine and central monoamine activity in fast and slow kindling rats. *Brain Research*, 840, 65–74.

McIntyre, D. C., Hutcheon, B., Schwabe, K., & Poulter, M. O. (2002). Divergent $GABA_A$ receptor-mediated synaptic transmission in genetically seizure-prone and seizure-resistant rats. *Journal of Neuroscience*, 22, 9922–9931.

MacKinnon, J. (1974). The behaviour and ecology of wild orangutans (*Pongo pygmaeus*). *Animal Behaviour*, **22**, 3–74.

McNamara, K. J. (1997). *Shapes of Time. The Evolution of Growth and Development.* The Johns Hopkins University Press: Baltimore, MD.

Maaswinkel, H., Baars, A. M., Gispen, W. H., & Spruijt, B. M. (1996). Roles of the basolateral amygdala and hippocampus in social recognition in rats. *Physiology & Behavior*, **60**, 55–63.

Maccoby, E. E. (1998). *The Two Sexes: Growing Up Apart, Coming Together.* Harvard University Press: Cambridge, MA.

Maccoby, E. E. & Jacklin, C. (1974). *The Psychology of Sex Differences.* Stanford University Press: Stanford, CA.

Maddison, W. P. & Maddison, D. R. (2000). *MacClade 4: Analysis of Phylogeny and Character Evolution.* Sinauer Associates: Sunderland, MA.

Markham, J. A., Morris, J. R., & Juraska, J. M. (2007). Neuron number decreases in the rat ventral, but not dorsal, medial prefrontal cortex between adolescence and adulthood. *Neuroscience*, **144**, 961–968.

Marler, P. (1987). Sensitive periods and the roles of specific and general sensory stimulation in bird-song learning. In J. P. Rauchecker & P. Marler (Eds.), *Imprinting and Cortical Plasticity* (pp. 99–135). Wiley & Sons: New York, NY.

Marler, P. (1991). Differences in behavioural development in closely related species: Birdsong. In P. Bateson (Ed.), *The Development and Integration of Behaviour. Essays in Honour of Robert Hinde* (pp. 41–70). Cambridge University Press: Cambridge, UK.

Martin, P. (1984a). The (four) whys and wherefores of play in cats: A review of functional, evolutionary, developmental, and causal issues. In P. K. Smith (Ed.), *Play in Animals and Humans* (pp. 71–94). Blackwell: Oxford, UK.

Martin, P. (1984b). The time and energy costs of play behaviour in the cat. *Zeitschrift für Tierpsychologie*, **64**, 298–312.

Martin, P. & Caro, T. (1985). On the function of play and its role in behavioral development. *Advances in the Study of Animal Behavior*, **15**, 59–103.

Masataka, N. (1994). Effects of experience with live insects on the development of fear of snakes in squirrel monkeys, *Saimiri sciureus. Animal Behaviour*, **46**, 741–746.

Mason, W. A. (1960). The effects of social restriction on the behavior of rhesus monkeys: I. Free social behavior. *Journal of Comparative & Physiological Psychology*, **53**, 582–589.

Mason, W. A. (1978). Social experience and primate cognitive development. In M. Bekoff & G. M. Burghardt (Eds.), *The Development of Behavior: Comparative and Evolutionary Aspects* (pp. 233–251). Garland STPM Press: New York, NY.

Mason, W. A. & Berkson, G. (1975). Effects of maternal mobility on the development of rocking and other behaviors in rhesus monkeys: A study with artificial mothers. *Developmental Psychobiology*, 8, 197–211.

Mather, J. A. & Anderson, R. C. (1999). Exploration, play, and habituation in octopuses (*Octopus doflieini*). *Journal of Comparative Psychology*, 113, 333–338.

May, C. J., Schank, J. G., Joshi, S., Tran, J., Taylor, R. J., & Scott, I.-E. (2006). Rat pups and random robots generate similar self-organized and intentional behavior. *Complexity*, 12, 53–66.

Meaney, M. J. (1988). The sexual differentiation of social play. *Trends in Neuroscience*, 11, 54–58.

Meaney, M. J. & Stewart, J. (1981). A descriptive study of social development in the rat (*Rattus norvegicus*). *Animal Behaviour*, 29, 34–45.

Meaney, M. J., Dodge, A. M., & Beatty, W. W. (1981). Sex-dependent effects of amygdaloid lesions on social play of prepubertal rats. *Physiology & Behavior*, 26, 467–472.

Meaney, M. J., Stewart, J., & Beatty, W. W. (1985). Sex differences in social play: The socialization of sex roles. *Advances in the Study of Behavior*, 15, 1–58.

Melvin, K. G., Doan, J., Pellis, S. M., Brown, L., Whishaw, I. Q., & Suchowersky, O. (2005). Pallidal deep brain stimulation and L-dopa do not improve qualitative aspects of skilled reaching in Parkinson's disease. *Behavioural Brain Research*, 160, 188–194.

Mendl, M. & Paul, E. S. (1990). Parental care, sibling relationships and the development of aggressive behaviour in two lines of wild house mice. *Behaviour*, 116, 11–37.

Merrick, N. J. (1977). Social grooming and play behavior of a captive group of chimpanzees. *Primates*, 18, 215–224.

Metz, G. A., Jadavji, N. M., & Smith, L. K. (2005). Modulation of motor function by stress: A novel concept of the effects of stress and corticosterone on behavior. *European Journal of Neuroscience*, 22, 1190–1120.

Metz, G. A., Schwab, M. E., & Welzl, H. (2001). The effects of acute and chronic stress on motor and sensory performance in male Lewis rats. *Physiology & Behavior*, 72, 29–35.

Metzgar, L. H. (1967). An experimental comparison of screech owl predation resident and transient white-footed mice (*Peromyscus leucopus*). *Journal of Mammalogy*, 48, 387–391.

Meunier, M., Bachevalier, J., Murray, E. A., Malkova, L., & Mishkin, M. (1999). Effects of aspiration versus neurotoxic lesions of the amygdala on emotional responses in monkeys. *European Journal of Neuroscience*, 11, 4403–4418.

Miller, S. (1981). Play. In D. McFarland (Ed.), *The Oxford Companion to Animal Behaviour*. Oxford University Press: Oxford, UK.

Miller, S. (1973). Ends, means and galumphing: Some leitmotifs of play. *American Anthropologist*, **75**, 87–98.

Miller, M. N. & Byers, J. A. (1998). Sparring as play in young pronghorn males. In M. Bekoff & J. A. Byers (Eds.), *Animal Play: Evolutionary, Comparative, and Ecological Perspectives* (pp.141–160). Cambridge University Press: Cambridge, UK.

Mills, M. G. L. (1990). *Kalahari Hyaena. Comparative Behavioural Biology of Two Species.* Unwin Hyman: London.

Mink, J. W. (1996). The basal ganglia: Focused selection and inhibition of competing motor programs. *Progress in Neurobiology*, **50**, 381–425.

Mitani, J. C. (1985). Mating behaviour of male orangutans in the Kutai Game Reserve, Indonesia. *Animal Behaviour*, **33**, 392–402.

Mitchell, G. (1979). *Behavioral Sex Differences in Nonhuman Primates.* Van Nostrand Reinhold: New York, NY.

Mohapel, P. & McIntyre, D. C. (1998). Amygdala kindling-resistant (SLOW) or –prone (FAST) rat strains show differential fear responses. *Behavioral Neuroscience*, **112**, 1402–1413.

Monnier, M. (1970). *Functions of the Nervous System: Vol. II. Motor and Sensorimotor Functions.* Elsevier: Amsterdam, The Netherlands.

Montessori, M. (1967). *The Absorbent Mind.* Holt, Rinehart & Winston: New York, NY. (Trans. Claremont, C. A.).

Mook, D. G. (1996). *Motivation. The Organization of Action*, 2nd Ed. W. W. Norton & Company: New York, NY.

Moore, C. L. (1985). Development of mammalian sexual behavior. In E. S. Gollin (Ed.), *The Comparative Development of Adaptive Skills* (pp. 19–56). Lawrence Erlbaum: Hillsdale, NJ.

Moore, M. M. (1985). Non-verbal courtship patterns in women: Contact and consequences. *Ethology & Sociobiology*, **6**, 237–247.

Moore, M. M. (1995). Courtship signaling and adolescents: "Girls just want to have fun"? *Journal of Sex Research*, **32**, 319–328.

Murie, J. O. & Michener, G. R. (1984). *The Biology of Ground-Dwelling Squirrels.* University of Nebraska Press: Lincoln, Nebraska.

Murphy, M. R., MacLean, P. D., & Hamilton, S. C. (1981). Species-typical behavior of hamsters deprived from birth of the neocortex. *Science*, **213**, 459–461.

Nekaris, A. & Bearder, S. K. (2006). The Lorisiform primates of Asia and Mainland Africa. In C. J. Campbell, A. Fuentes, K. C. MacKinnon, M. Panger & S. K. Bearder (Eds.), *Primates in Perspective* (pp. 24–45). Oxford University Press: Oxford, UK.

Nelson, R. J. (2000). *An Introduction to Behavioral Endocrinology.* 2nd Edn. Sunderland, Sinauer Associates: MA, USA.

Neuringer, A. (2004). Reinforced variability in animals and people: Implications for adaptive action. *American Psychologist,* **59,** 891–906.

Newell, T. G. (1971). Social encounters in two prosimian species: *Galago crassicaudatus* and *Nycticebus coucang. Psychonomic Society,* **2,** 128–130.

Niesink, R. J. M. & van Ree, J.M. (1982). Short-term isolation increases social interactions of male rats: A parametric analysis. *Physiology & Behavior,* **29,** 819–825.

Nowak, R. W. (1999). *Walker's Mammals of the World.* 6th Edn. Vol. II. The Johns Hopkins University Press: Baltimore, MD.

Nunes, S., Muecke, E.-M., Anthony, J. A., & Batterbee, A. S. (1999). Endocrine and energetic mediation of play behavior in free-living Belding's ground squirrels. *Hormones & Behavior,* **36,** 153–165.

Nunes, S., Muecke, E.-M., Sanchez, Z., Hoffmeier, R. R., & Lancaster, L. T. (2004a). Play behavior and motor development in juvenile Belding's ground squirrels (*Spermophilus beldingi*). *Behavioral Ecology & Sociobiology,* **56,** 97–105.

Nunes, S., Lancaster, L. T., Miller, N. A., Mueller, M. A., Muelhaus, J., & Castro, L. (2004b). Functions and consequences of play behaviour in juvenile Belding's ground squirrels. *Animal Behaviour,* **68,** 27–37.

Ogawa, T., Mikuni, M., Kuroda, Y., Muneoka, K., Mori, K. J., & Takahashi, K. (1994). Periodic maternal deprivation alters stress response in adult offspring, potentiates the negative feedback regulation of restraint-induced adrenocortical response, and reduces the frequencies of open field-induced behaviors. *Pharmacology, Biochemistry & Behavior,* **49,** 961–967.

Ohman, A. (2002). Automaticity and the amygdala: Nonconcious responses to emotional faces. *Current Directions in Psychological Science,* **11,** 62–66.

Olds, J. & Milner, P. (1954). Positive reinforcement produced by electrical stimulation of septal area and other regions of the rat brain. *Journal of Comparative & Physiological Psychology,* **47,** 419–427.

Olomon, C. M., Breed, M. D., & Bell, W. J. (1976). Ontogenetic and temporal aspects of agonistic behavior in a cockroach, *Periplanata americana. Behavioral Biology,* **17,** 243–248.

Orgeur, P. & Signoret, J. P. (1984). Sexual play and its functional significance in the domestic sheep (*Ovis aries* L.). *Physiology & Behavior,* **33,** 111–118.

Ortega, J. C. & Bekoff, M. (1987). Avian play: Comparative, evolutionary, and developmental trends. *Auk,* **104,** 338–341.

Oyama, S. (2000). *The Ontogeny of Information: Developmental Systems and Evolution.* 2nd Edn. Duke University Press: Durham, NC.

Pagel, M. D. & Harvey, P. H. (1993). Evolution of the juvenile period in mammals. In M. E. Pereira & L. A. Fairbanks (Eds.), *Juvenile Primates: Life History, Development and Behavior* (pp. 28–37). Oxford University Press: Oxford, UK.

Paglieri, F. (2005). Playing by and with the rules: Norms and morality in play development. *Topoi*, **24**, 149–167.

Palagi, E. (2006). Social play in bonobos (*Pan paniscus*) and chimpanzees (*Pan troglodytes*): Implications for natural social systems and interindividual relationships. *American Journal of Physical Anthropology*, **129**, 418–426.

Palagi, E. (2008). Sharing the motivation to play: The use of signals in adult bonobos. *Animal Behaviour*, **75**, 887–896.

Palagi, E. (in press). Adult play fighting in a prosimian (*Lemur catta*): Modalities and roles of tail signals. *Journal of Comparative Psychology*, in press.

Palagi, E. & Paoli, T. (2007). Play in adult bonobos (*Pan paniscus*): Modality and potential meaning. *American Journal of Physical Anthropology*, **134**, 219–225.

Palagi, E., Cordoni, G., & Borgognini Tarli, S. M. (2004). Immediate and delayed benefits of play behavior: New evidence from chimpanzees (*Pan troglodytes*). *Ethology*, **110**, 949–962.

Palagi, E., Paoli, T., & Borgognini Tarli, S. (2006). Short-term benefits of play behavior and conflict prevention in *Pan paniscus*. *International Journal of Primatology*, **27**, 1257–1270.

Panksepp, J. (1981). The ontogeny of play in rats. *Developmental Psychobiology*, **14**, 327–332.

Panksepp, J. (1998). *Affective Neuroscience.* Oxford University Press: Oxford, UK.

Panksepp, J. (2007). Neuroevolutionary sources of laughter and social joy: Modeling primal human laughter in laboratory rats. *Behavioural Brain Research*, **182**, 231–244.

Panksepp, J. & Beatty, W. W. (1980). Social deprivation and play in rats. *Behavioral & Neural Biology*, **30**, 197–206.

Panksepp, J. & Burgdorf, J. (2000). 50-kHz chirping laughter in response to conditioned and unconditioned tickle-induced reward in rats: Effects of social housing and genetic variables. *Behavioural Brain Research*, **115**, 25–38.

Panksepp, J. & Burgdorf, J. (2003). "Laughing" rats and the evolutionary antecedents of human joy? *Physiology & Behavior*, **79**, 533–547.

Panksepp, J., Normansell, L., Cox, J. F., & Siviy, S. M. (1994). Effects of neonatal decortication on the social play of juvenile rats. *Physiology & Behavior*, **56**, 429–443.

Panksepp, J., Siviy, S. M., & Normansell, L. (1984). The psychobiology of play: Theoretical and methodological perspectives. *Neuroscience & Biobehavioral Reviews*, **8**, 465–492.

Panksepp, J. B., Jochem, K. A., Kim, J. U., Koy, J. J., Wilson, E. D., Chen, Q., Wilson, C. R., & Lahvis, G. P. (2007). Affiliative behavior, ultrasonic communication and social reward are influenced by genetic variation in adolescent mice. *PLoS ONE*, **4**, e351.

Panksepp, J. B. & Lahvis, G. P. (2007). Social reward among juvenile mice. *Genes, Brain & Behavior*, **6**, 661–671.

Paquette, D. (1994). Fighting and play fighting in captive adolescent chimpanzees. *Aggressive Behavior*, **20**, 49–65.

Paquette, D. (2004). Theorizing the father-child relationship: Mechanisms and developmental outcomes. *Human Development*, **47**, 193–219.

Paquette, D., Carbonneau, R., Dubeau, D., Bigras M., & Tremblay, R. (2003). Prevalence of father-child rough-and-tumble play and physical aggression in preschool children. *European Journal of Psychology & Education*, **18**, 171–189.

Parke, R. D., Cassidy, J., Burks, V. S., Carson, J., & Boyum, L. (1992). Familial contributions to peer competence among young children: The role of interactive and affective processes. In R. D. Parke & C. Ladd (Eds.), *Family-Peer Relationships* (pp. 107–134). Erlbaum: Hillsdale, NJ.

Parker, S. T. & McKinney, M. L. (1999). *Origins of Intelligence: The Evolution of Cognitive Development in Monkeys, Apes, and Humans.* Johns Hopkins University Press: Baltimore, MD.

Pasztor, T. J., Smith, L. K., MacDonald, N. L., Michener, G. R., & Pellis, S. M. (2001). Sexual and aggressive play fighting of sibling Richardson's ground squirrels. *Aggressive Behavior*, **27**, 323–337.

Pederson, J. M., Glickman, S. E., Frank, L. G., & Beach, F. A. (1990). Sex differences in the play behavior of immature spotted hyenas, *Crocuta crocuta. Hormones & Behavior*, **24**, 403–420.

Pellegrini, A. D. (1988). Elementary school children's rough-and-tumble play and social competence. *Developmental Psychology*, **24**, 802–806.

Pellegrini, A. D. (1992). Rough-and-tumble play and social problem solving flexibility. *Creativity Research Journal*, **5**, 13–27.

Pellegrini, A. D. (1995a). Boys' rough-and-tumble play and social competence: Contemporaneous and longitudinal relations. In: A. D. Pellegrini (Ed.), *The Future of Play Theory: A Multidisciplinary Inquiry into the Contribution of Brian Sutton-Smith* (pp. 107–126). State University of New York Press: Albany: NY.

Pellegrini, A. D. (1995b). Boys' rough-and-tumble play, social competence and group composition. *British Journal of Developmental Psychology*, **11**, 237–248.

Pellegrini, A. D. (2003). Perceptions and possible functions of play and real fighting in early adolescence. *Child Development*, **74**, 1552–1533.

Pellegrini, A. D. (2006). The development and function of rough-and-tumble play in childhood and adolescence: A sexual selection theory perspective. In A. Göncü & S. Gaskins (Eds.), *Play and Development* (pp. 77–98). Lawrence Erlbaum Associates: Mahwah, NJ.

Pellegrini, A. D. & Archer, J. (2005). Sex differences in competitive and aggressive behavior: A view from sexual selection theory. In B. J. Ellis & D. J. Bjorklund (Eds.), *Origins of the Social Mind: Evolutionary Psychology and Child Development* (pp. 219–244). Guilford Press: New York, NY.

Pellegrini, A. D. & Bjorklund, D. F. (1997). The role of recess in children's cognitive performance. *Educational Psychologist*, **32**, 35–40.

Pellegrini, A. D. & Long, J. D. (2002). A longitudinal study of bullying, dominance, and victimization during the transition from primary to secondary school. *British Journal of Developmental Psychology*, **20**, 259–280.

Pellegrini, A. D. & Smith, P. K. (1998). Physical activity play: The nature and function of a neglected aspect of play. *Child Development*, **69**, 577–598.

Pellegrini, A. D. & Smith, P. K., (2005). *The Nature of Play. Great Apes and Humans.* Guilford Press: New York, NY.

Pellegrini, A. D., Dupuis, D., & Smith, P. K. (2006). Play in evolution and development. *Developmental Review*, **27**, 261–276.

Pellegrini, A. D., Horvat, M., & Huberty, P. (1998). The relative cost of children's physical play. *Animal Behaviour*, **55**, 1053–1061.

Pellegrini, A. D., Long, J. D., Roseth, C., Bohn, K., & van Ryzin, M. (2007). A short-term longitudinal study of preschoolers' (*Homo sapiens*) sex segregation: The role of physical activity, sex, and time. *Journal of Comparative Psychology*, **121**, 282–289.

Pellis, S. M. (1981a). Exploration and play in the behavioural development of the Australian magpie *Gymnorhina tibicen*. *Bird Behaviour*, **3**, 37–49.

Pellis, S. M. (1981b). A description of social play by the Australian magpie *Gymnorhina tibicen* based on Eshkol-Wachman notation. *Bird Behaviour*, **3**, 61–79.

Pellis, S. M. (1983). Development of head and foot coordination in the Australian Magpie *Gymnorhina tibicen*, and the function of play. *Bird Behaviour*, **4**, 57–62.

Pellis, S. M. (1984). Two aspects of play-fighting in a captive group of Oriental small-clawed otters *Amblonyx cinerea*. *Zeitschrift für Tierpsychologie*, **65**, 77–83.

Pellis, S. M. (1988). Agonistic versus amicable targets of attack and defense: Consequences for the origin, function and descriptive classification of play-fighting. *Aggressive Behavior*, **14**, 85–104.

Pellis, S. M. (1989). Fighting: The problem of selecting appropriate behavior patterns. In R. J. Blanchard, P. F. Brain, D. C. Blanchard & S. Parmigiani (Eds.), *Ethoexperimental Approaches to the Study of Behavior* (pp. 361–374). Kluwer Academic Publishers: Dordrecht, The Netherlands.

Pellis, S. M. (1993). Sex and the evolution of play fighting: A review and a model based on the behavior of muroid rodents. *The Journal of Play Theory & Research*, **1**, 56–77.

Pellis, S. M. (1996). Righting and the modular organization of motor programs. In K.-P. Ossenkopp, M. Kavaliers & P. R. Sanberg (Eds.), *Measuring Movement and Locomotion: From Invertebrates to Humans* (pp. 115–133). Landes Company: Austin, TX.

Pellis, S. M. (1997). Targets and tactics: The analysis of moment-to-moment decision making in animal combat. *Aggressive Behavior*, **23**, 107–129.

Pellis, S. M. (2002a). Keeping in touch: Play fighting and social knowledge. In M. Bekoff, C. Allen & G. M. Burghardt (Eds.), *The Cognitive Animal: Empirical and Theoretical Perspectives on Animal Cognition* (pp. 421–427). MIT Press: Cambridge, MA.

Pellis, S. M. (2002b). Sex-differences in play fighting revisited: Traditional and non-traditional mechanisms for sexual differentiation in rats. *Archives of Sexual Behavior*, **31**, 11–20.

Pellis, S. M. (2005). Cross-species comparisons. In B. Hopkins (Ed.), *The Cambridge Encyclopedia of Child Development* (pp. 112–114). Cambridge University Press: Cambridge, UK.

Pellis, S. M. & Iwaniuk, A. N. (1999a). The roles of phylogeny and sociality in the evolution of social play in muroid rodents. *Animal Behaviour*, **58**, 361–373.

Pellis, S. M. & Iwaniuk, A. N. (1999b). The problem of adult play: A comparative analysis of play and courtship in primates. *Ethology*, **105**, 783–806.

Pellis, S. M. & Iwaniuk, A. N. (2000a). Comparative analyses of the role of postnatal development on the expression of play fighting. *Developmental Psychobiology*, **36**, 136–147.

Pellis, S. M. & Iwaniuk, A. N. (2000b). Adult-adult play in primates: Comparative analyses of its origin, distribution and evolution. *Ethology*, **106**, 1083–1104.

Pellis, S. M. & Iwaniuk, A. N. (2002). Brain system size and adult-adult play in primates: A comparative analysis of the roles of the non-visual neocortex and the amygdala. *Behavioural Brain Research*, **134**, 31–39.

Pellis, S. M. & Iwaniuk, A. N. (2004). Evolving a playful brain: A levels of control approach. *International Journal of Comparative Psychology*, **17**, 90–116.

Pellis, S. M. & McKenna, M. M. (1992). Intrinsic and extrinsic influences on play fighting in rats: Effects of dominance, partner's playfulness, temperament and neonatal exposure to testosterone propionate. *Behavioural Brain Research*, **50**, 135–145.

Pellis, S. M. & McKenna, M. M. (1995). What do rats find rewarding in play fighting? An analysis using drug-induced non-playful partners. *Behavioural Brain Research*, **68**, 65–73.

Pellis, S. M. & Pasztor, T. J. (1999). The developmental onset of a rudimentary form of play fighting in mice, *Mus musculus*. *Developmental Psychobiology*, **34**, 175–182.

Pellis, S. M. & Pellis, V. C. (1982). Do post-hatching factors limit clutch size in the Cape Barren goose (*Cereopsis novaehollandiae* Latham)? *Australian Wildlife Research*, **9**, 145–149.

Pellis, S. M. & Pellis, V. C. (1983). Locomotor-rotational movements in the ontogeny and play of the laboratory rat *Rattus norvegicus. Developmental Psychobiology*, **16**, 269–286.

Pellis, S. M. & Pellis, V. C. (1987). Play-fighting differs from serious fighting in both target of attack and tactics of fighting in the laboratory rat *Rattus norvegicus. Aggressive Behavior*, **13**, 227–242.

Pellis, S. M. & Pellis, V. C. (1988a). Play-fighting in the Syrian golden hamster *Mesocricetus auratus* Waterhouse, and its relationship to serious fighting during post-weaning development. *Developmental Psychobiology*, **21**, 323–337.

Pellis, S. M. & Pellis, V. C. (1988b). Identification of the possible origin of the body target which differentiates play-fighting from serious fighting in Syrian golden hamsters *Mesocricetus auratus. Aggressive Behavior*, **14**, 437–449.

Pellis, S. M. & Pellis, V. C. (1989). Targets of attack and defense in the play fighting by the Djungarian hamster *Phodopus campbelli*: Links to fighting and sex. *Aggressive Behavior*, **15**, 217–234.

Pellis, S. M. & Pellis, V. C. (1990). Differential rates of attack, defense and counterattack during the developmental decrease in play fighting by male and female rats. *Developmental Psychobiology*, **23**, 215–231.

Pellis, S. M. & Pellis, V. C. (1991a). Role reversal changes during the ontogeny of play fighting in male rats: Attack versus defense. *Aggressive Behavior*, **17**, 179–189.

Pellis, S. M. & Pellis, V. C. (1991b). Attack and defense during play fighting appear to be motivationally independent behaviors in muroid rodents. *The Psychological Record*, **41**, 175–184.

Pellis, S. M. & Pellis, V. C. (1992a). Juvenilized play fighting in subordinate male rats. *Aggressive Behavior*, **18**, 449–457.

Pellis, S. M. & Pellis, V. C. (1992b). An analysis of the targets and tactics of conspecific attack and predatory attack in Northern grasshopper mice *Onychomys leucogaster. Aggressive Behavior*, **18**, 301–316.

Pellis, S. M. & Pellis, V. C. (1993). The influence of dominance on the development of play fighting in pairs of male Syrian golden hamsters (*Mesocricetus auratus). Aggressive Behavior*, **19**, 293–302.

Pellis, S. M. & Pellis, V. C. (1996). On knowing it's only play: The role of play signals in play fighting. *Aggression & Violent Behavior*, **1**, 249–268.

Pellis, S. M. & Pellis, V. C. (1997a). Targets, tactics and the open mouth face during play fighting in three species of primates. *Aggressive Behavior*, **23**, 41–57.

Pellis, S. M. & Pellis, V. C. (1997b). The pre-juvenile onset of play fighting in rats (*Rattus norvegicus*). *Developmental Psychobiology*, 31,193–205.

Pellis, S. M. & Pellis, V. C. (1998a). The play fighting of rats in comparative perspective: A schema for neurobehavioral analyses. *Neuroscience & Biobehavioral Reviews*, 23, 87–101.

Pellis, S. M. & Pellis, V. C. (1998b). Structure-function interface in the analysis of play. In M. Bekoff & J. A. Byers (Eds.), *Animal Play: Evolutionary, Comparative, and Ecological Perspectives* (pp.115–140). Cambridge University Press: Cambridge, UK.

Pellis, S. M. & Pellis, V. C. (2006). Play and the development of social engagement: A comparative perspective. In P. J. Marshall & N. A. Fox (Eds.), *The Development of Social Engagement: Neurobiological Perspectives* (pp. 247–274). Oxford University Press: Oxford, UK.

Pellis, S. M. & Pellis, V. C. (2007). Rough-and-tumble play and the development of the social brain. *Current Directions in Psychological Science*, 16, 95–98.

Pellis, S. M., Field, E. F., & Whishaw, I. Q. (1999). The development of a sex-differentiated defensive motor-pattern in rats: A possible role for juvenile experience. *Developmental Psychobiology*, 35, 156–164.

Pellis, S. M., Pellis, V. C., & Dewsbury, D. A. (1989). Different levels of complexity in the playfighting by muroid rodents appear to result from different levels of intensity of attack and defense. *Aggressive Behavior*, 15, 297–310.

Pellis, S. M., Pellis, V. C., & Foroud, A. (2005). Play fighting: Aggression, affiliation and the development of nuanced social skills. In R. Tremblay, W. W. Hartup & J. Archer (Eds.), *Developmental Origins of Aggression* (pp. 47–62). Guilford Press: New York.

Pellis, S. M., Pellis, V. C., & Kolb, B. (1992). Neonatal testosterone augmentation increases juvenile play fighting but does not influence adult dominance relationships in male rats. *Aggressive Behavior*, 18, 437–447.

Pellis, S. M., Pellis, V. C., & McKenna, M. M. (1993). Some subordinates are more equal than others: Play fighting amongst adult subordinate male rats. *Aggressive Behavior*, 19, 385–393.

Pellis, S. M., Pellis, V. C., & McKenna, M. M. (1994). A feminine dimension in the play fighting of rats (*Rattus norvegicus*) and its defeminization neonatally by androgens. *Journal of Comparative Psychology*, 108, 68–73.

Pellis, S. M., Pellis, V. C., & Nelson, J. E. (1992). The development of righting reflexes in the pouch young of the marsupial *Dasyurus hallucatus*. *Developmental Psychobiology*, 25, 105–125.

Pellis, S. M., Pellis, V. C., & Reinhart, C. J. (in press). The evolution of social play. In: C. Worthman, P. Plotsky & D. Schechter (Eds.), *Formative Experiences: The*

Interaction of Caregiving, Culture, and Developmental Psychobiology. Cambridge University Press: Cambridge, UK.

Pellis, S. M., Pellis, V. C., & Teitelbaum, P. (1991). Air-righting without the cervical righting reflex in adult rats. *Behavioural Brain Research*, 45, 185–188.

Pellis, S. M., Pellis, V. C., & Whishaw, I. Q. (1992). The role of the cortex in play fighting by rats: Developmental and evolutionary implications. *Brain, Behavior & Evolution*, 39, 270–284.

Pellis, S. M., O'Brien, D. P., Pellis, V. C., Teitelbaum, P., Wolgin, D. L., & Kennedy, S. (1988). Escalation of feline predation along a gradient from avoidance through "play" to killing. *Behavioral Neuroscience*, 102, 760–777.

Pellis, S. M., Pellis, V. C., Pierce, J. D., Jr., & Dewsbury, D. A. (1992). Disentangling the contribution of the attacker from that of the defender in the differences in the intraspecific fighting of two species of voles. *Aggressive Behavior*, 18, 425–435.

Pellis, S. M., Castañeda, E., McKenna, M. M., Tran-Nguyen, L. T. L., & Whishaw, I. Q. (1993). The role of the striatum in organizing sequences of play fighting in neonatally dopamine-depleted rats. *Neuroscience Letters*, 158, 13–15.

Pellis, S. M., Field, E. F., Smith, L. K., & Pellis, V. C. (1997). Multiple differences in the play fighting of male and female rats. Implications for the causes and functions of play. *Neuroscience & Biobehavioral Reviews*, 21, 105–120.

Pellis, S. M., Pasztor, T. J., Pellis, V. C., & Dewsbury, D. A. (2000). The organization of play fighting in the grasshopper mouse (*Onychomys leucogaster*): Mixing predatory and sociosexual targets and tactics. *Aggressive Behavior*, 26, 319–334.

Pellis, S. M., Hastings, E., Shimizu, T., Kamitakahara, H., Komorowska, J., Forgie M. L., & Kolb, B. (2006). The effects of orbital frontal cortex damage on the modulation of defensive responses by rats in playful and non-playful social contexts. *Behavioral Neuroscience*, 120, 72–84.

Pellis, V. C., Pellis, S. M., & Teitelbaum, P. (1991). A descriptive analysis of the postnatal development of contact-righting in rats (*Rattus norvegicus*). *Developmental Psychobiology*, 24, 237–263.

Pereira, M. E. & Altmann, J. (1985). Development of social behavior in free-living nonhuman primates. In E. S. Watts (Ed.), *Nonhuman Primate Models for Human Growth and Development* (pp. 217–309). Alan R. Liss: New York, NY.

Peterson, J. B. & Flanders, J. L. (2005). Play and the regulation of aggression. In R. E. Tremblay, W. W. Hartup & J. Archer (Eds.), *Developmental Origins of Aggression* (pp. 133–157). Guilford Press: New York, NY.

Petrù, M., Spinka, M., Lhota, S., & Sípek, P. (2008). Head rotations in the play of Hanuman langurs (*Semnopithecus entellus*): A description and an analysis of function. *Journal of Comparative Psychology*, 122, 9–18.

Pfeifer, R. & Bongard, J. (2007). *How the Body Shapes the Way We Think. A New View of Intelligence.* MIT Press: Cambridge, MA.

Pierce, J. D., Jr., Pellis, V. C., Dewsbury, D. A., & Pellis, S. M. (1991). Targets and tactics of agonistic and precopulatory behavior in montane and prairie voles: Their relationship to juvenile play fighting. *Aggressive Behavior*, **17**, 337–349.

Pinel, J. P. J. & Treit, D. (1978). Burying as a defensive response in rats. *Journal of Comparative & Physiological Psychology*, **92**, 708–71.

Pinel, J. P. J., Symons, L. A., Christensen, B. K., & Tees, R. C. (1989). Development of defensive burying in *Rattus norvegicus*. Experience and defensive responses. *Journal of Comparative Psychology*, **103**, 359–365.

Pinker, S. (1994). *The Language Instinct.* William Morrow: New York, NY.

Pionanotti, M. R. & Vieira, M. L. (2004). Presence of the father and parental experience have differentiated effects on pup development in Mongolian gerbils (*Meriones unguiculatus*). *Behavioural Processes*, **66**, 107–117.

Plato (1975). *The Laws.* Penguin: Harmondsworth, Middlesex, UK. (Trans. T. J. Saunders).

Plotsky, P. M. & Meaney, M. J. (1999). Early, postnatal experience alters hypothalamic corticotropin-releasing factor (CRF) mRNA, median eminence CRF content and stress-induced release in adult rats. *Molecular Brain Research*, **19**, 195–200.

Poole, T. B. & Fish, J. (1975). An investigation of playful behaviour in *Rattus norvegicus* and *Mus musculus* (Mammalia). *Journal of Zoology*, **175**, 61–71.

Potegal, M. & Einon, D. (1989). Aggressive behaviors in adult rats deprived of play fighting experience as juveniles. *Developmental Psychobiology*, **22**, 159–172.

Power, T. G. (1985). Mother- and father-infant play: A developmental analysis. *Child Development*, **56**, 1514–1524.

Power, T. G. (2000). *Play and Exploration in Animals and Children.* Lawrence Erlbaum Associates: Mahwah, NJ.

Power, T. G. & Parke, R. D. (1983). Patterns of mother and father play with their 8-month old infants: A multiple analysis approach. *Infant Behavior & Development*, **6**, 453–459.

Powers, W. T. (1973). *Behavior: The Control of Perception.* Wildwood House: London, UK.

Pozis-Francois, O., Zahavi. A., & Zahavi, A. (1999). Social play in Arabian babblers. *Behaviour*, **141**, 425–450.

Prather, M. D., Lavenex, P., Mauldin-Joundain, M. L., Mason, W. A., Capitanio, J. P., Mendoza, S. P., & Amaral, D. G. (2001). Increased social fear and decreased fear of objects in monkeys with neonatal amygdala lesions. *Neuroscience*, **106**, 653–658.

Priest, R. F. & Swain, J. E. (2002). Humor and its implications for leadership effectiveness. *Humor*, **15**, 169–189.

Provine, R. R. (2000). *Laughter: A Scientific Investigation.* Viking: New York, NY.

Prusky, G. T., Harker, K. T., Douglas, R. M., & Whishaw, I. Q. (2002). Variation in visual acuity within pigmented, and between pigmented and albino rat strains. *Behavioural Brain Research,* **136,** 339–348.

Pryce, C. R. & Feldon, J. (2003). Long-term neurobehavioural impact of the postnatal environment in rats: Manipulations, effects and mediating mechanisms. *Neuroscience & Biobehavioral Reviews,* **27,** 57–71.

Pryce, C. R., Bettschen, D., & Feldon, J. (2001). Comparison of the effects of early handling and early deprivation on maternal care in the rat. *Developmental Psychobiology,* **38,** 239–251.

Reeve, H. K. & Sherman, P. W. (1993). Adaptation and the goals of evolutionary research. *The Quarterly Review of Biology,* **68,** 1–32.

Reinhart, C. J., Pellis, S. M., & McIntyre, D. C. (2004). The development of play fighting in kindling-prone (FAST) and kindling–resistant (SLOW) rats: How does the retention of phenotypic juvenility affect the complexity of play? *Developmental Psychobiology,* **45,** 83–92.

Reinhart, C. J., Metz, G., Pellis, S. M., & McIntyre, D. C. (2006). Play fighting between kindling-prone (FAST) and kindling-resistant (SLOW) rats. *Journal of Comparative Psychology,* **120,** 19–30.

Renner, M. J. & Rosenzweig, M. R. (1986). Social interactions among rats housed in grouped and enriched conditions. *Developmental Psychobiology,* **19,** 303–313.

Riess, B. F. (1954). Effect of altered environment and of age in the mother-young relationships among animals. *Annals of the New York Academy of Sciences,* **57,** 606–610.

Robbins, T. W., Jones, G. H., & Wilkinson, L. S. (1966). Behavioural and neurochemical effects of early social deprivation in the rat. *Journal of Psychopharmacology,* **10,** 39–47.

Robinson, T. E. & Berridge, K. C. (1993). The neural basis of drug craving: An incentive-sensitization theory of addiction. *Brain Research Review,* **18,** 247–91.

Robitaille, J. A. & Bovet, J. (1976). Field observations on the social behaviour of the Norway rat, *Rattus norvegicus* (Berkenhout). *Biology of Behaviour,* **1,** 289–308.

Romeo, R. D., Karatsoreos, I. N., & McEwen, B. S. (2006). Pubertal maturation and time of day differentially affect behavioral and neuroendocrine responses following an acute stressor. *Hormones & Behavior,* **50,** 463–468.

Romer, A. S. (1970). *The Vertebrate Body.* 4th Edn. W. B. Saunders Company: Philadelphia, PA.

Roozendaal, B. (2002). Stress and memory: Opposing effects of glucocorticoids on memory consolidation and memory retrieval. *Neurobiology of Learning & Memory,* **78,** 578–595.

Rose, M. R. & Lauder, G. V. (1996). *Adaptation.* Academic Press: San Diego, CA.

Rosenzweig, M. R. & Bennett, E. L. (1972). Cerebral changes in rats exposed individually to an enriched environment. *Journal of Comparative & Physiological Psychology*, **80**, 304–313.

Ruedi-Betschen, D., Pedersen, E. M., Feldon, J., & Pryce, C. R. (2005). Early deprivation under specific conditions leads to reduced interest in reward in adulthood in Wistar rats. *Behavioural Brain Research*, **156**, 297–310.

Ruse, M. (2003). *Darwin & Design. Does Evolution have a Purpose?* Harvard University Press: Cambridge, MA.

Rushen, J. & Pajor, E. (1987). Offence and defense in fights between young pigs (*Sus scrofa*). *Aggressive Behavior*, **13**, 329–346.

Russon, A. E., Vasey, P. L., & Gauthier, C. (2002). Seeing in the mind's eye: Eye-covering play in orangutans and Japanese macaques. In R. W. Mitchell (Ed.), *Pretending and Imagination in Animals and Children* (pp. 241–254). Cambridge University Press: Cambridge, UK.

Ryan, K. M. & Mohr, S. (2005). Gender differences in playful aggression during courtship in college students. *Sex Roles*, **53**, 591–601.

Sapolsky, R. M. (1995). *Why Zebras Don't Get Ulcers: A Guide to Stress, Stress-Related Diseases, and Coping.* Freeman: New York, NY.

Scarlett, W. G., Naudeau, S., Salonius-Pasternak, D., & Ponte, I., (2005). *Children's Play.* Sage Publications: Thousand Oaks, CA.

Schino, G., Scucchi, S., Maestripieri, D., & Turillazzi, P. G. (1988). Allogrooming as a tension-reduction mechanism: A behavioral approach. *American Journal of Primatology*, **16**, 43–60.

Schmidt-Nielsen, K. (1984). *Scaling. Why is Animal Size so Important?* Cambridge University Press: Cambridge, UK.

Schneider, M. & Koch, M. (2005). Deficient social and play behavior in juvenile and adult rats after neocortical lesion: Effects of chronic pubertal cannabinoid treatment. *Neuropharmacology*, **30**, 944–957.

Schurmann, C. (1982). Mating behavior of wild orangutans. In: L. E. M. De Boer (Ed.), *The Orang Utan. Its Biology and Conservation* (pp. 269–289). Dr. W. Junk Publishers: The Hague, The Netherlands.

Seward, J. P. (1945). Aggressive behavior in the rat: I. General characteristics: Age and sex differences. *Journal of Comparative Psychology*, **38**, 175–197.

Sgofio, A., Stili, D., Musso, E., Mainardi, D., & Parmigiani, S. (1992). Offensive and defensive bite-target topographies in attacks by lactating rats. *Aggressive Behavior*, **17**, 47–52.

Shakespeare, W. (1952). Hamlet. In W. G. Clarke & W. A. Wright (Eds.), *The Plays and Sonnets of William Shakespeare*, Vol. II. Encyclopaedia Britannica: London, UK.

Sharpe, L. L. (2005a). Play fighting does not affect subsequent fighting success in wild meerkats. *Animal Behaviour,* **69**, 1023–1029.

Sharpe, L. L. (2005b). Play does not enhance social cohesion in a cooperative mammal. *Animal Behaviour,* **70**, 551–558.

Sharpe, L. L. (2005c). Frequency of social play does not affect dispersal partnerships in wild meerkats. *Animal Behaviour,* **70**, 559–569.

Sharpe, L. L. & Cherry, M. I. (2003). Social play does not reduce aggression in wild meerkats. *Animal Behaviour,* **66**, 989–997.

Siegel, A. (2004). *Neurobiology of Aggression and Rage.* CRC Press: Boca Raton, FL.

Simpson, M. J. A. (1976). The study of animal play. In P. P. G. Bateson & R. A. Hinde (Eds.), *Growing Points in Ethology* (pp. 385–400). Cambridge University Press: Cambridge, UK.

Siviy, S. M. (1998). Neurobiological substrates of play behavior: Glimpses into the structure and function of mammalian playfulness. In M. Bekoff & J. A. Byers (Eds.), *Animal Play: Evolutionary, Comparative, and Ecological Perspectives* (pp. 221–242). Cambridge University Press: Cambridge, UK.

Siviy, S. M. & Atrens, D. M. (1992). The energetic costs of rough-and-tumble play in the juvenile rat. *Developmental Psychobiology,* **25**, 137–148.

Siviy, S. M. & Harrison, K. A. (2008). Effects of neonatal handling on play behavior and fear towards a predator odor in juvenile rats (*Rattus norvegicus*). *Journal of Comparative Psychology,* **122**, 1–8.

Siviy, S. M. & Panksepp, J. (1987). Sensory modulation of juvenile play in rats. *Developmental Psychobiology,* **20**, 39–55.

Siviy, S. M., Baliko, C. N., & Bowers, K. S. (1997). Rough-and-tumble play behavior in Fischer-344 and buffalo rats: Effects of social isolation. *Physiology & Behavior,* **61**, 597–602.

Siviy, S. M., Harrison, K.A., & McGregor, L. S. (2006). Fear, risk assessment, and playfulness in the juvenile rat. *Behavioral Neuroscience,* **120**, 49–59.

Siviy, S. M., Fleischhauer, A. E., Kerrigan, L. A., & Kuhlman, S. J. (1996). D2 dopamine receptor involvement in the rough-and-tumble play behavior of juvenile rats. *Behavioral Neuroscience,* **110**, 1168–1176.

Siviy, S. M., Love, N. J., DeCicco, B. M., Giordano, S. B., & Seifert, T. L. (2003). The relative playfulness of juvenile Lewis and Fischer-344 rats. *Physiology & Behavior,* **80**, 385–394.

Smith, E. O. (1978). A historic view on the study of play. Statement on the problem. In E. O. Smith (Ed.), *Social Play in Primates* (pp. 1–32). Academic Press: New York, NY.

Smith, L. K. & Metz, G. A. (2005). Dietary restrictions alters fine motor function in rats. *Physiology & Behavior,* **85**, 581–592.

Smith, L. K., Fantella, S.-L., & Pellis, S. M. (1999). Playful defensive responses in adult male rats depend upon the status of the unfamiliar opponent. *Aggressive Behavior*, **25**, 141–152.

Smith, L. K., Forgie, M. L., & Pellis, S. M. (1998a). Mechanisms underlying the absence of the pubertal shift in the playful defense of female rats. *Developmental Psychobiology*, **33**, 147–156.

Smith, L. K., Forgie, M. L., & Pellis, S. M. (1998b). The post-pubertal change in the playful defense of male rats depends upon neonatal exposure to gonadal hormones. *Physiology & Behavior*, **63**, 151–155.

Smith, L. K., Field, E. F., Forgie, M. L., & Pellis, S. M. (1996). Dominance and age-related changes in the play fighting of intact and post-weaning castrated males (*Rattus norvegicus*). *Aggressive Behavior*, **22**, 215–226.

Smith, P. K. (1982). Does play matter? Functional and evolutionary play. *Behavioral Brain Sciences*, **5**, 139–184.

Smith, P. K. (1997). Play fighting and real fighting. Perspectives on their relationship. In Schmitt, A., Atzwanger, K., Grammar, K., & Schäfer, K. (Eds.), *New Aspects of Human Ethology* (pp. 47–64). Plenum Press: New York, NY.

Smith, P. K. (2005). Play: Types and functions in human development. In B. J. Ellis & D. F. Bjorklund (Eds.), *Origins of the Social Mind* (pp. 271–291). Guilford Press: New York, NY.

Smith, P. K. & Boulton, M. (1990). Rough and tumble play, aggression and dominance: Perception and behavior in children's encounters. *Human Development*, **33**, 271–282.

Smith, P. K., Smees, R., & Pellegrini, A. D. (2004). Play fighting and real fighting: Using video playback methodology with young children. *Aggressive Behavior*, **30**, 164–173.

Smith, P. K., Hunter, T., Carvalho, A. M. A., & Costabile, A. (1992). Children's perceptions of playfighting, playchasing and real fighting: A cross-national interview study. *Social Development*, **1**, 211–229.

Snyder, R. J., Zhang, A. J., Zhang, Z. H., Li, G. H., Tian, Y. Z., Huang, X. M., Luo, L., Bloomsmith, M. A., Forthman, D. L., & Maple, T. L. (2003). Behavioral and developmental consequences of early rearing experience for captive giant pandas (*Ailuropoda melanoleuca*). *Journal of Comparative Psychology*, **117**, 235–245.

Spinka, M., Newberry, R. C., & Bekoff, M. (2001). Mammalian play: Can training for the unexpected be fun? *Quarterly Review of Biology*, **76**, 141–176.

Stone, A. I. (2008). Seasonal effects on play behavior in immature *Saimiri sciureus* in Eastern Amazonia. *International Journal of Primatology*, **29**, 195–205.

Striedter, G. F. (2005). *Principles of Brain Evolution*. Sinauer Associates: Sunderland, MA.

Suchecki, D., Nelson, D. Y., van Oers, H., & Levine, S. (1995). Activation and inhibition of the hypothalamic-pituitary-adrenal axis of the neonatal rat: Effects of maternal deprivation. *Psychoneuroendocrinology,* **20,** 169–182.

Suomi, S. J. (2005). Genetic and environmental factors influencing the expression of impulsive aggression and serotonergic functioning in rhesus monkeys. In R. E. Tremblay, W. W. Hartup & J. Archer (Eds.), *Developmental Origins of Aggression* (pp. 63–82). Guilford Press: New York, NY.

Sutton-Smith, B. (1982). *A History of Children's Play: The New Zealand Playground, 1840–1950.* University of Pennsylvania Press: Philadelphia, PA.

Sutton-Smith, B. (1997). *The Ambiguity of Play.* Harvard University Press: Cambridge, MA.

Sutton-Smith, B. & Kelly-Byrne, D. (1984). The idealization of play. In P. K. Smith (Ed.), *Play in Animals and Humans* (pp. 305–321). Blackwell: Oxford, UK.

Symons, D. (1978). *Play and Aggression. A Study of Rhesus Monkeys.* Columbia University Press: New York, NY.

Tamis-LeMonda, C. S. (2004). Conceptualizing fathers' roles: Playmates and more. *Human Development,* **47,** 220–227.

Tamis-LeMonda, C. S., Shannon, J. D., Cabrera, N. J., & Lamb, M. F. (2004). Fathers and mothers at play with their 2- and 3-year-olds: Contributions to language and cognitive development. *Child Development,* **75,** 1806–1820.

Teitelbaum, P. & Pellis, S. M. (1992). Towards a synthetic physiological psychology. *Psychological Science,* **3,** 4–20.

Terranova, M. L., Laviola, G., & Alleva, E. (1993). Ontogeny of amicable social behavior in the mouse: Gender differences and ongoing isolation outcomes. *Developmental Psychobiology,* **26,** 467–481.

Terranova, M. L., Laviola, G., de Acestis, L., & Alleva, E. (1998). A description of the ontogeny of agonistic behavior of mice. *Journal of Comparative Psychology,* **112,** 3–12.

Thierry, B. (2004). Social epigenesis. In: B. Thierry, M. Singh & W. Kaumanns (Eds.), *Macaque Societies. A Model for the Study of Social Organization* (pp. 267–290). Cambridge University Press: Cambridge, UK.

Thierry, B. (2005). Integrating proximate and ultimate causation: Just one more go! *Current Science,* **89,** 1180–1183.

Thierry, B. (2007). Behaviorology divided: Shall we continue? *Behaviour,* **144,** 861–878.

Thierry, B., Iwaniuk, A. N., & Pellis, S. M. (2000). The influence of phylogeny on the social behaviour of macaques (Primates: Cercopithecidae, genus *Macaca*). *Ethology,* **106,** 713–728.

Thompson, K. V. (1998). Self assessment in juvenile play. In M. Bekoff & J. A. Byers

(Eds.), *Animal Play: Evolutionary, Comparative, and Ecological Perspectives* (pp.183–204). Cambridge University Press: Cambridge, UK.

Thor, D. H. & Holloway Jr., W. R., Jr. (1983a). Scopolamine blocks play fighting behavior in juvenile rats. *Physiology & Behavior*, 30, 545–549.

Thor, D. H. & Holloway, W. R., Jr. (1983b). Play solicitation behavior in juvenile male and female rats. *Animal Learning & Behavior*, 11, 173–178.

Thor, D. H. & Holloway, W. R., Jr. (1984a). Social play in juvenile rats: A decade of methodological and experimental research. *Neuroscience & Biobehavioral Reviews*, 8, 455–464.

Thor, D. H. & Holloway, W. R., Jr. (1984b). Developmental analysis of social play behavior in juvenile rats. *Bulletin of the Psychonomic Society*, 22, 587–590.

Timmann, D., Richter, S., Schoch, B., & Frings, M. (2006). Cerebellum and cognition: A review of the literature. *Aktuelle Neurologie*, 33, 70–80.

Tremblay, R. E. & Nagin, D. S. (2005). The developmental origins of human aggression in humans. In R. E. Tremblay, W. W. Hartup & J. Archer (Eds.), *Developmental Origins of Aggression* (pp. 83–105). Guilford Press: New York, NY.

Trezza, V. & Vanderschuren, L. J. M. J. (2008). Bidirectional canabinoid modulation of social behavior in adolescent rats. *Psychopharmacology*, 197, 217–227.

Trut, L. N. (1999). Early canid domestication: The fox-farm experiment. *American Scientist*, 87, 160–169.

Uylings, H. B. M., van Eden, C. G., Parnavelas, J. G., & Kalsbeek, A. (1990). The prenatal and postnatal development of rat cerebral cortex. In B. Kolb & R. C. Tees (Eds.), *The Cerebral Cortex of the Rat* (pp. 35–76). MIT Press: Cambridge, MA.

van den Berg, C. L., Hol, T., van Ree, J. M., Spruijt, B. M., Everts, H., & Koolhaas, J. M. (1999). Play is indispensable for an adequate development of coping with social challenges in the rat. *Developmental Psychobiology*, 34, 129–138.

Vanderschuren, L. J. M. J., Niesink, R. J. M., & van Ree, J. M. (1997). The neurobiology of play behavior in rats. *Neuroscience & Biobehavioral Reviews*, 21, 309–326.

van Oers, H. J. J., de Kloet, E. R., Li, C., & Levine, S. (1998). The ontogeny of glucocorticoid negative feedback: Influence of maternal deprivation. *Endocrinology*, 139, 2838–2846.

van Oortmerssen, G. A. (1971). Biological significance, genetics, and evolutionary origin of variability in behavior within and between inbred strains of mice (*Mus musculus*). *Behaviour*, 38, 1–91.

van Schaik, C. (2004). *Among Orangutans. Red Apes and the Rise of Human Culture.* Harvard University Press: Cambridge, MA.

Varlinskaya, E. I. & Spear, L. P. (2008). Social interactions in adolescent and adult Sprague-Dawley rats: Impact of social deprivation and test context familiarity. *Behavioural Brain Research*, 188, 398–405.

Varlinskaya, E. I., Spear, L. P., & Spear, N. E. (1999). Social behavior and social motivation in adolescent rats: Role of housing conditions and partner's activity. *Physiology & Behavior*, **67**, 475–482.

von Frijtag, J. C., Schot, M., van den Bos, R., & Spruijt, B. M. (2002). Individual housing during the play period results in changed responses to and consequences of a psychosocial stress situation in rats. *Developmental Psychobiology*, **41**, 58–69.

Walf, A. A. & Frye, C. A. (2007). The use of the elevated plus maze as an assay of anxiety-related behavior in rodents. *Nature Protocols*, **2**, 322–328.

Walker, C. & Byers, J. A. (1991). Heritability of locomotor play in house mice, *Mus domesticus*. *Animal Behaviour*, **42**, 891–897.

West, M. J. & King, A. P. (1987). Settling nature and nurture into an ontogenetic niche. *Developmental Psychobiology*, **10**, 549–562.

West-Eberhardt, M. J. (2003). *Developmental Plasticity*. Oxford University Press: New York, NY.

Whishaw, I. Q. (1988). Food wrenching and dodging: Use of an action pattern for the analysis of sensorimotor and social behavior in the rat. *Journal of Neuroscience Methods*, **24**, 169–178.

Whishaw, I. Q. & Kolb, B. (1985). The mating movements of male decorticate rats: Evidence for subcortically generated movements by the male but regulation of approaches by the female. *Behavioural Brain Research*, **17**, 171–191.

Whishaw, I. Q. & Kolb, B. (Eds.) (2004). *The Behavior of the Laboratory Rat*. Oxford University Press: Oxford, UK.

Whishaw, I. Q. & Pellis, S. M. (1990). The structure of skilled forelimb reaching in the rat: A proximally driven movement with a distal rotatory component. *Behavioural Brain Research*, **41**, 49–59.

Whishaw, I. Q., Sarna, J., & Pellis, S. M. (1998). Evidence for rodent-common and species-typical limb and digit use in eating derived from a comparative analysis of ten rodent species. *Behavioural Brain Research*, **96**, 79–91.

Whishaw, I. Q., Pellis, S. M., Gorny, B. P., & Pellis, V. C. (1991). The impairments in reaching and the movements of compensation in rats with motor cortex lesions: An endpoint, videorecording, and movement notation analysis. *Behavioural Brain Research*, **42**, 77–91.

Whishaw, I. Q., Metz, G., Kolb, B., & Pellis, S. M. (2001). Accelerated nervous system development contributes to behavioral efficiency in the laboratory mouse: A behavioral review and theoretical proposal. *Developmental Psychobiology*, **39**, 151–170.

Whishaw, I. Q., Suchowersky, O., Davis, L., Sarna, J., Metz, G. A., & Pellis, S. M. (2002). Impairment of pronation, supination, and body co-ordination in reach-to-grasp tasks in human Parkinson's disease (PD) reveals homology to deficits in animal models. *Behavioural Brain Research*, **133**, 165–176.

Whishaw, I. Q., Gorny, B., Foroud, A., & Kleim, J. A. (2003). Long-Evans and Sprague-Dawley rats have similar skilled reaching success and limb representations in motor cortex but different movements: Some cautionary insights into the selection of rat strains for neurobiological motor research. *Behavioural Brain Research*, **145**, 221–232.

White, S. W., Koenig, K., & Scahill, L. (2007). Social skills development in children with autism spectrum disorders: A review of the intervention literature. *Journal of Autism & Developmental Disorders*, **37**, 1858–1868.

Williams, G. C. (1966). *Adaptation and Natural Selection*. Princeton University Press: Princeton, NJ.

Williams, G. C. (1992). *Natural Selection: Domains, Levels and Challenges*. Oxford University Press: New York, NY.

Wilmer, A. H. (1991). Behavioral deficiencies of aggressive 8–9-year-old boys: An observational study. *Aggressive Behavior*, **17**, 135–154.

Wilson, M. L. (2005). *An Investigation into the Factors that Affect Play Fighting Behavior in Giant Pandas*. Unpublished Ph.D thesis. Georgia Institute of Technology: Atlanta, GA.

Wilson, S. (1973). The development of social behaviour in the vole (*Microtus agrestis*). *Zoological Journal of the Linnean Society*, **52**, 45–62.

Wolff, R. J. (1981). Solitary and social play in wild *Mus musculus* (Mammalia). *Journal of Zoology*, **195**, 405–412.

Wolfheim, J. H. (1977). Sex differences in behavior in a group of juvenile talapoin monkeys (*Miopithecus talapoin*). *Behaviour*, **63**, 110–128.

Wolgin, D. L. (1982). Motivation, activation, and behavioral integration. In R. L. Isaacson & N. E. Spear (Eds.), *The Expression of Knowledge* (pp. 243–290). Plenum Press: New York, NY.

Wright, I. K., Upton, N., & Marden, C. A. (1991). Resocialization of isolation-reared rats does not alter their anxiogenic profile on the elevated X-maze model of anxiety. *Physiology & Behavior*, **50**, 1129–1132.

Young, L. J., Lim, M. M., Gingrich, B., & Insel, T. R. (2001). Cellular mechanisms of social attachment. *Hormones & Behavior*, **40**, 133–138.

Zahr, L. K. & Balian, S. (1995). Responses of premature infants to routine nursing interventions and noise in the NICU. *Nursing Research*, **41**, 179–185.

Zucker, E. L. & Clarke, M. R. (1992). Development and comparative aspects of social play in mantled howling monkeys in Costa Rica. *Behaviour*, **123**, 144–171.

INDEX